MOLECULAR BIOLOGY AND GENETIC ENGINEERING

Prof. N.K. Yadava

MANGLAM PUBLICATIONS
DELHI-110053 (INDIA)

Published by :
MANGLAM PUBLICATIONS
L-21/1, Street No. 5, Shivaji Marg, Near Kali Mandir,
J.P. Nagar, Kartar Nagar, West Ghonda, Delhi-110053
Phone : 011-22945677, (M.) 9968367559, 9868572512
E-mail: manglam.books2007@rediffmail.com
manglam.publications2007@rediffmail.com

Molecular Biology and Genetic Engineering

First Edition 2009

ISBN 978-93-80013-19-0

Price Rs. 850/-

PRINTED IN INDIA

Published by Ravindra Kumar Yadav for Manglam Publications, Delhi-110053 Laser Typeset by M.K. Computers, Delhi- 53, Printed at Sachin Printers, Moujpur Delhi- 53

PREFACE

The present title *Molecular Biology and Genetic Engineering* is one of the youngest branches of Biology. It has expanded so rapidly in its few decade of existence there is a wealth of material to draw on the ever accumulating information on the subject has been judiciously tailored, lucidly presented and profusely illustrated to meet the needs of the under-graduate and post-graduate students. The title is not intended to be comprehensive, more could it be at this length but concentrates as putting across the basic principles of the subject as briefly and lucidly as possible. It does this with the aid of careful selected examples, some recent and other classical of the field with numerous illustrations. The aim is to familiarize the readers with the active and exciting area of research and to lays a solid foundation on which further study of its various facts may be based.

The author expresses his thanks to all those friends, colleagues, and research scholars whose continuous inspirations have initiated him to bring this title.

The author wishes to thank the Mr. D.P. Yadav, Director of M/s. Manglam Publications, Delhi for bringing out this book.

Constructive criticisms and suggestions for improvement of the book will be thankfully acknowledged

N. K. Yadava

Contents

1

DNA Structure

DNA is central to the chemistry of life; its involvement extends beyond the storage and duplication of genetic information to include participation in the regulation of its own expression and evolutionary development. Interaction with the many other cellular components involved in these processes depends essentially on nonbonded interactions, and these are modulated by the spatial relationships of atoms imposed by the three-dimensional structure of the macromolecule.

The conformational study of the DNA molecule provides an opportunity to observe structural detail at a level often not accessible in models that are more functionally complete but unworkably complex. Investigations on simple analogous systems are potentially, therefore, keys to understanding, in molecular terms, the operation of these important biological mechanisms.

1.1 TECHNIQUES

Natural DNA is classically a high polymer of four mononucleotide units. The superficial regularity of this arrangement in many cases hinders conformational study at the level of resolution required to describe the properties of relatively short sequences of specialized function.

Chemical synthesis of oligonucleotides has in recent years made an important contribution here, enabling pioneering work on DNA isolatedd from natural sources to be extended to fragments of known sequence. Wide-ranging physicochemical and biochemical techniques have been applied to provide, in varying degrees of completeness, different and often complementary views of DNA structure.

This is true not only of the parameters that are accessible by different methods, but also of the restrictions in molecular environment imposed by individual experimental systems.

In vivo, DNA forms only one component of' the apparatus employed in its storage and use: it is associated with proteins, other nucleic acids, and species of low molecular weight, all of which can be expected to influence its structure in some degree. Such interactions are paralleled by those in solutions or crystals whose contents are designed to model the biological environment.

1.1.1 X-Ray Diffraction

Native DNA precipitates from aqueous solutions as a gel that can he drawn into fibers. Individual molecules are aligned both with the fiber axis and in register with each other according to the regularly repeating structure of the sugar-phosphate backbone.

Xray diffraction patterns of such fibers reflect the periodicity of regions of high electron density such as base planes viewed edgeon, or the crossing points of the backbone. Although these patterns alone are inadequate to define the three-dimensional structure in detail, they can provide constraints from which to deduce generalized models such as the familiar A and B forms of DNA.

Single crystals are essential for high resolution structural studies. Fragments of natural or enzymatically polymerized DNA produced by mechanical shearing and size fractionation can form single crystals, but they are insufficiently ordered for structure determination at high resolution.

Synthetic oligonucleotides, by contrast, commonly form highly ordered single crystals that diffract at or near atomic resolution (2.5-0.9Å). The molecular structure can be derived directly from the experimental data using mathematical methods at high resolution (up to 1.3Å). Otherwise the structure can be solved only if a suitable starting model is found which can be refined to fit the experimental data to a reliability factor (R) of 15-25%.

In such an analysis the conformational details, albeit not the individual atomic positions, are accurately determined. The X-ray data also contain information about intermolecular interactions, either between DNA oligomers themselves, or with proteins and drugs in mixed crystals, all of which are valuable in understanding in vivo behaviour.

Oligonucleotide crystals contain approximately 50% by weight of water. Only a small proportion of these water molecules are sufficiently ordered for their positions in the crystal to he defined precisely, but those close to available H-bond donors and acceptors in the nucleic acid can be identified, providing information about the contribution of hydration to the stability of secondary structure.

The chief limitation of the method is the need for single crystals. The chances of crystallizing a random sequence are small but are improved by the selection of sequences compatible with known packing motifs.

At a practical level, already determined structures are used as templates whose limited modification can allow other sequences to be examined, or the effects of modified nucleotides or noncomplementary bases observed. A few structures have also been solved that are not recognizable as fragments of a conventional extended duplex.

1.1.2 NMR Spectroscopy

When studying DNA in solution, nuclear magnetic resonance spectroscopy is the method of choice for obtaining comprehensive structural information. The method takes advantage of the abundance of hydrogen atoms in nucleotides: diverse magnetic environments both within the individual monomer units and those superimposed by the secondary structure cause the proton spectra often to he well dispersed. Conformational mobility of oligonucleotides in solution can result in broadening of spectral lines, but a range of spectroscopic data is accessible when signals are sharp.

In particular, three-bond spin-spin couplings and the rates of nuclear Overhauser effect buildup between protons provide information on torsion angles and interproton distances, respectively. These data can be used as constraints to derive reliable computer-generated three-dimensional structures. Ring current effects of adjacent heterocyclic bases within the secondary structure give rise to substantial variation in the chemical shifts of signals of otherwise equivalent protons.

Duplex formation is associated with generally upfield shifts in the resonances of base protons, when increased stacking places the protons close to the faces of adjacent aromatic rings. Precise structural interpretation is not possible, but distinctively shifted signals are indicators of structural anomaly. Present capabilities owe much to the development of two-dimensional acquisition techniques, which greatly enhance the effective spectral resolution.

In favourable cases data obtainable from spectrometers operating in the range of 400-600 MHz for protons can define the conformation of an oligonucleotide containing up to 15 unique nucleotide residues. Some specific conformational investigations are possible on longer sequences. Only a few milligrams of sample is required at these magnetic field strengths.

1.1.3 Vibrational Spectra

Bands in infrared and Raman spectra correspond to differences between the vibrational energy levels of bonds. An infrared absorption line is observed if the relevant bending or stretching mode is associated with a change in dipole moment, while for Raman scattering a change in polarizability is required.

The spectra may be seen as complementary; but in practice, for complex molecules such as DNA, reality lies between these extremes, and infrared and Raman spectra are often qualitatively similar. Individual lines in the spectra, though they may be due predominantly to the behaviour of one particular bond, are combination bands of all deformations in the vicinity and so contain conformational information.

They refer necessarily to localized features; however, some of these have in turn been shown to correlate with global properties of the double helix and so are of more general significance as indicators of three-dimensional structure. Vibrational frequencies such as those due to carbonyl stretching are directly interpretable-for example, in terms of bond strength changes due to hydrogen bonding; but in other cases only an empirical approach is possible.

By comparing spectra of polynucleotides with known and different conformations, investigators have identified groups of bands associated with specific structural features-for example, in distinguishing *2'-endo* and *3'-endo-fu*ranose conformations typical of B- and A-DNA, respectively, and from which more general conclusions can be drawn.

A particular advantage of the method is its applicability to a range of solid and solution samples, including mixtures. It enables the comparison of structures in a number of physical environments, including those for which no other methods for extracting structural information are available.

1.1.4 Electronic Spectra

Nucleic acid bases possess strong ultraviolet chromophores with extinction coefficients of around 10'. Spectra of oligo- and polynucleotides resemble the sum of their component nucleotides, but the molecular environment within primary, secondary, and tertiary structures, where the bases are in close proximity. modifies the resultant spectra.

The hypochromic effect observed at wavelengths close to the absorption maximum (ca. 260 nm) upon association of complementary strands into a duplex is commonly applied to monitor the temperature-dependent helix-random coil transition, or melting of DNA. At a more detailed level, changes in the relative intensities of component bands of the absorption envelope cause its shape to change as a result of secondary structure in the polymer.

The chiral environments of secondary structures, such as the various families of double helix, give rise to distinctive circular dichroism (CD) spectra. The potential for obtaining structural information combined with its sensitivity make CD a simple and attractive technique for

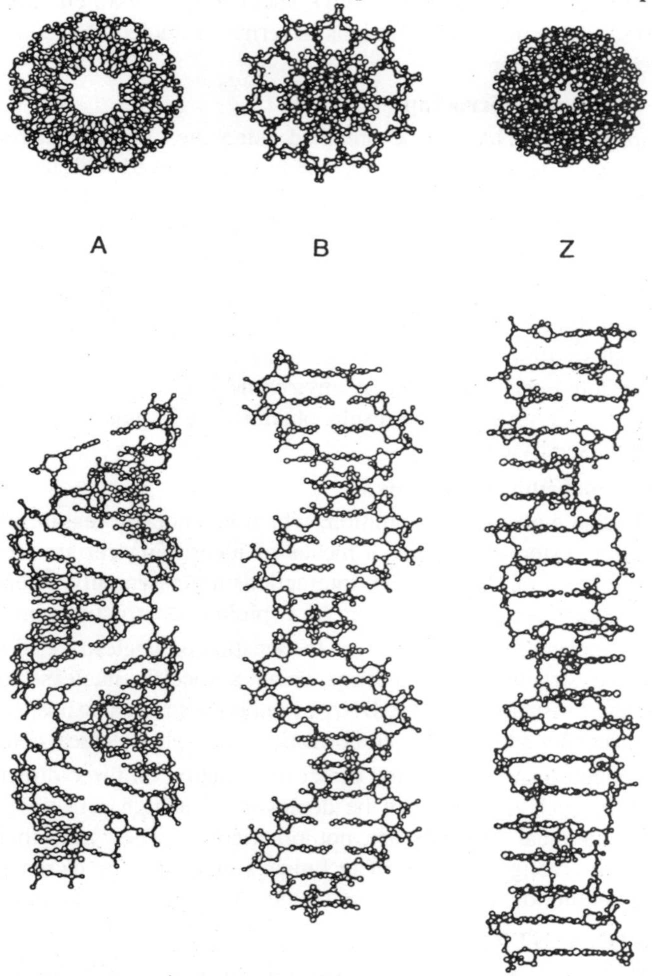

Figure 1.1: Skeletal representations of 20 base pair fragments of A-, B-, and Z-DNA according to idealized coordinates derived from X-ray diffraction data obtained from fibers (A- and B-) and single crystals of d(CG)3 (Z-).

conformational study. It is, however, a comparative method that can detect changes in conformation or identifies a stereotype by reference to standard spectra.

Structures must first be identified using X-ray diffraction or some other definitive technique. The observation that A, B, and Z forms of DNA produce characteristic CD spectra has enabled extensive investigation of their interrelations in terms of nucleotide sequence or solution composition.

1.1.5 Electron Microscopy

Images of DNA can be obtained using the electron microscope. Early studies, which required specimens to be treated with heavy atom stains, contributed much to our knowledge of higher order DNA structure and topology. More recently, developments in scanning tunneling microscopy have made possible the visualization of the general form of the double helix such that its component residues can be discerned and its dimensions and periodicity measured.

With so alluringly visual a technique, it is especially important to bear in mind the experimental environment and sample preparation involved. Reassuringly, the results obtained are in general accord with the established view.

1.1.6 Biochemical Methods

Precise molecular recognition, which is a notable feature of many biological systems, provides a means of identifying certain structural characteristics of DNA. When interactions are essentially independent of nucleotide sequence, they can be exploited as structural probes.

Examples of this approach are the location of singlestranded regions in looped structures by cleavage using a specific nuclease and the histological visualization of Z-DNA by antibodies raised against synthetic polynucleotides adopting that conformation under physiological conditions.

Though providing limited direct information on three-dimensional structure, such methods can be deployed with high sensitivity when physicochemical methods are not practicable. They are, therefore, important among the array of techniques available for increasing our understanding of the complex behaviour of DNA in living things.

1.2 DNA STRUCTURE

Fiber diffraction patterns defined the classical double helix. These sequence-averaged data imply a necessarily regular structure within which the base pairs are interchangeable. High resolution studies, chiefly using single crystal X-ray analysis, have penetrated this superficial regularity to expose sequence-linked variations consistent with the inherent flexibility of the sugar-phosphate backbone.

The A, B, and Z forms of DNA are well established. Early studies showed that DNA-B was generally associated with higher water activity

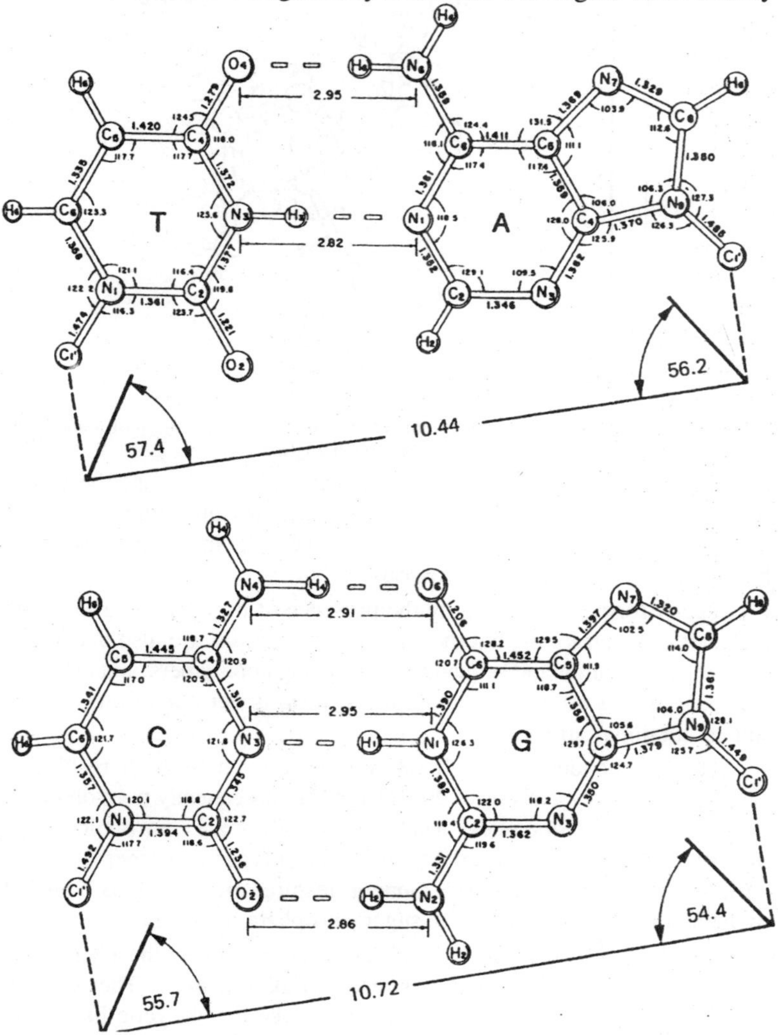

Figure 1.2: Watson-Crick base pairs formed in the accurately determined X-ray crystal structures of r(ApU) and r(GpC), which illustrate the geometry of the glycosidic bonds and other characteristic dimensions applicable to analogous pairs in DNA..

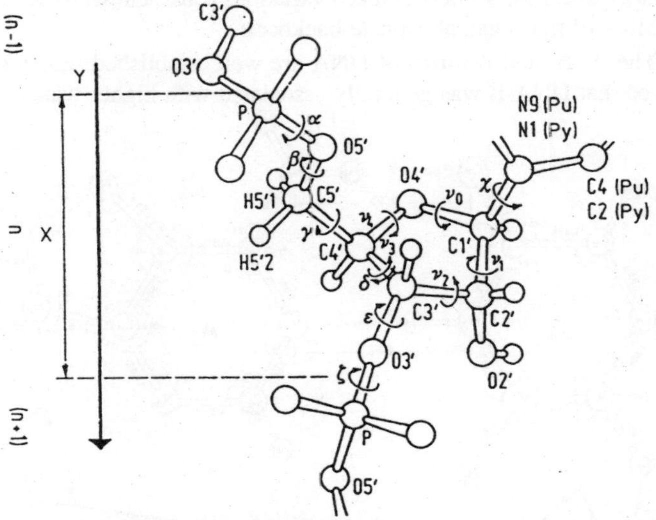

Figure 1.3: A deoxyribose-phosphate unit in the backbone, showing atom numberings and the notation used to identify backbone torsion angles.

and lower ionic strength than the A form. DNA-B was, therefore, presumed to he prevalent in living systems. Results with fibers are paralleled in solutions of polynucleotides where the interconversions can be monitored using CD spectroscopy.

The A form is stabilized by the presence of organic cosolvents and high salt concentrations, but the transition conditions are also dependent on base composition and sequence. Alternating copolymers, notably poly [d(C-G)] and its analogues, adopt the Z form under conditions of high salt or torsional stress.

This left-handed double helix was discovered by high resolution single crystal X-ray analysis of d(CG), but has subsequently been observed by a variety of techniques. The three families are morphologically distinct, though within especially the A and B types a number of minor variations have been described. General features are two clefts bounded by helical ridges formed by the sugar-phosphate backbone.

These grooves, denoted "major" and "minor," reflect the asymmetric disposition of glycosidic bonds in Watson-Crick base pairs. They contain the hydrogen bond donors and acceptors of the heterocyclic bases and are sources of recognition information, since the arrangement of functional groups is unique to the particular nucleotide sequence.

Groove geometry is important to the accessibility of these groups and varies greatly between the different families, while the relative

Table 1.1: Typical parameters associated with the three DNA families.

Parameter	*A-DNA*	*B-DNA*	*Z-DNA*
Helix sense	Right handed	Right handed	Left handed
Base pairs per turn	II	10	12
Helical twist	33°	36°	10°, -50°
Rise per base pair	2.9Å	3.4Å	3.7Å
Helix pitch	32Å	34Å	35Å
Tilt	13°	0°	7
Roll°	6.0 1 4.7°	- I ± 5.5°	3.4 ± 2.1°
Glycosidic conformation	*anti-*	*anti-*	*anti, syn*
Propeller twist	15.4 ± 6.2°	11.7 1 4.8°	4.4 ± 2.8°

disposition of the features depends on other conformational characteristics of the double helix. In B-DNA the base pairs are roughly perpendicular to and intersect with the helix axis, giving an approximately 10-fold repeat. With this arrangement the major and minor grooves are respectively 10 and 5 A wide and have similar depths.

The A form appears hollow when viewed end-on. The base pairs are displaced from the helix axis and inclined to give a repeat of about 11. Though of similar widths to those of the B form, the minor groove is shallower and the major groove much deeper. Z-DNA is a lefthanded double helix with the base pairs perpendicular to the helix axis.

The helical repeat is approximately 12 base pairs but, in contrast to the right-handed forms, individual nucleotide units are not interchangeable. The true repeating unit is a dinucleotide, since the glycosidic conformations of successive nucleotides are alternately syn- and anti-, and in consequence the backbone follows a zigzag course. Slight displacement of the base pairs from the center of the helix produces a major groove that is very shallow, but the minor groove is deep and extends to the helix axis.

The Z conformation is principally associated with alternating purinepyrimidine sequences in which the purine has a syn-glycosidic conformation. Characteristic dimensions of the helical types are listed in Table elsewhere in this chapter.

Key conformational features have been rigorously defined according to the schemes in Figures elsewhere in this chapter. The three families of duplex DNA are the result of subtle equilibria between cohesive and

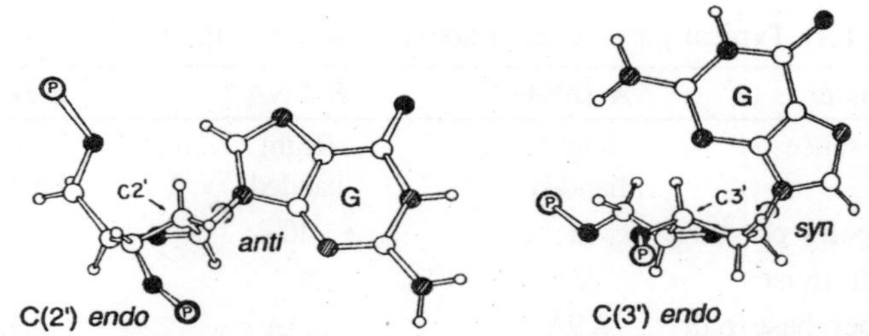

Figure 1.4: Principal conformations of nucleoside units in DNA illustrated by guanosine in B-DNA and Z-DNA, respectively; both the sugar pucker and orientation about the glycosidic bond are close to the extreme values for these parameters.

disruptive forces within the secondary structure and its envelope of solvent molecules and counterions. These include stacking between the heterocyclic bases, hydrogen bonds, and ionic interactions around the phosphodiester groups and other centers of localized charge.

Within basic structural outlines, variations are seen that may serve to shape molecular surfaces suited to specific recognition events. Correlations between certain backbone torsion angles are observed experimentally and appear to result from accommodation that preserves the overall helical shape.

Too few structures have, however, been deter-mined firmly to establish more comprehensive descriptions of sequence dependence. Unless balanced by the presence of positive ions, repulsion between the numerous centers of negative charge, principally phosphate and car-bonyl oxygen atoms, is a major obstacle to formation of multiple-stran-ded helices. This is also the case for higher order structure.

In vivo, these cations are provided by proteins: histones in eukaryotic chromatin form a core on which the double helix is wrapped before further compaction by subsequent levels of coiling.

In the absence of proteins, DNA condenses into higher order structures when polyamines such as spermine are added. The intermolecular interactions in these assemblies may be similar to those in oligonucleotide crystals, where close contacts between duplexes are observed.

Certain DNA sequences can associate into three- and fourstranded assemblies using the abundance of hydrogen-bonding sites of the four DNA bases. Triplexes composed of purine·pyrimidine·purine (A·T·A, G·C·G) and pyrimidine·purine·pyrimidine (T·A·T, C·G·CH+) base triples have been characterized by chemical, biochemical, and spectroscopic techniques that establish the direction of the third strand bound in the

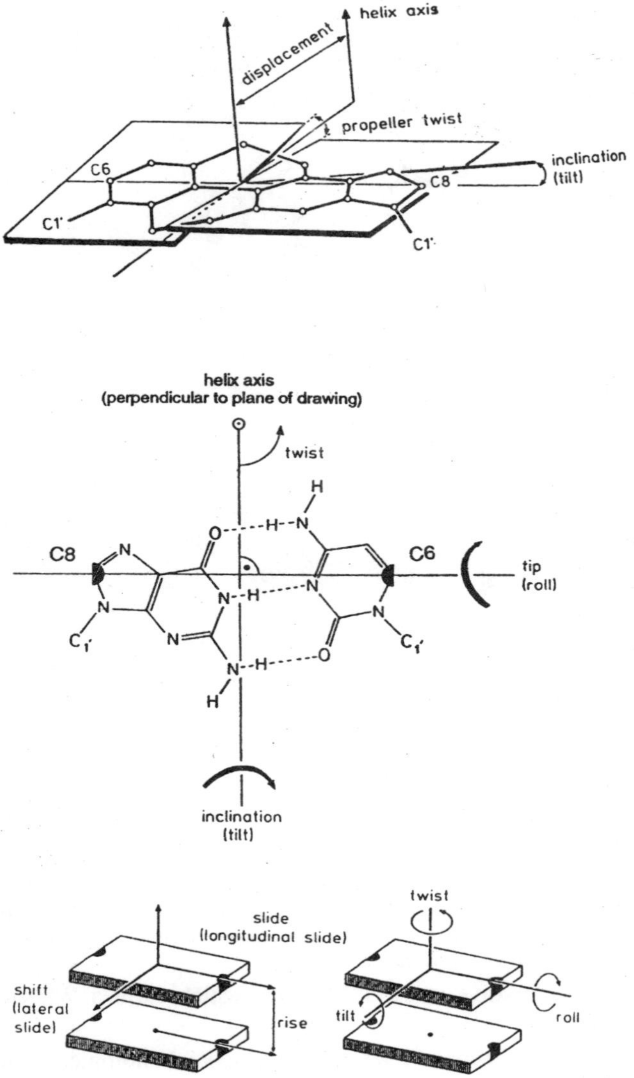

Figure 1.5: Schematic definitions of the numerous parameters devised to describe the geometry of individual base pairs and the relationships between successive steps of the double helix. Alternatives are given in parentheses.

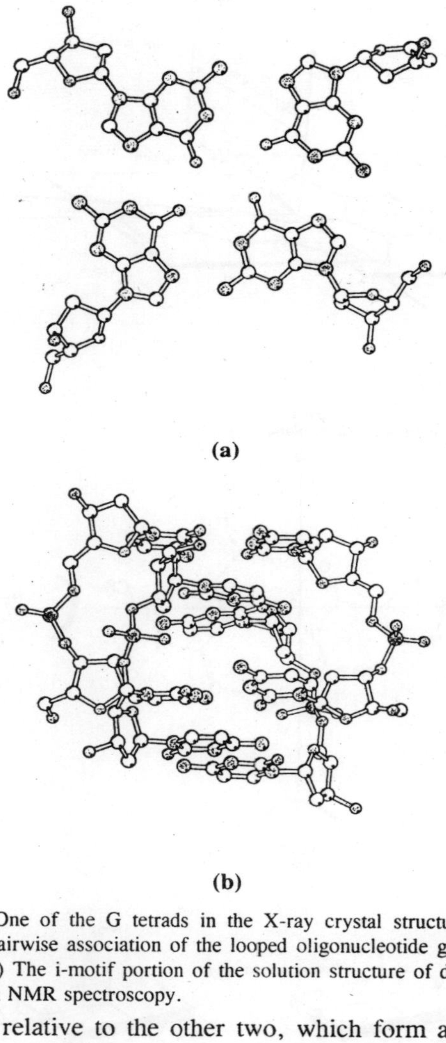

(a)

(b)

Figure 1.6: (a) One of the G tetrads in the X-ray crystal structure d(GGGGTTTT GGGG), where pairwise association of the looped oligonucleotide generates fragments of quadruplex. (b) The i-motif portion of the solution structure of d(TCC) determined by high resolution NMR spectroscopy.

major groove relative to the other two, which form a Watson-Crick double helix.

Particular sequences may relate to special biological functions. Characteristic guanine repeats are found, for example, in telomeres at the ends of eukaryotic chromosomes. Fragments of related sequence have been shown to associate into quadruplex structures made up of guanine tetrads and have been extensively studied by NMR and X-ray analysis.

Deoxycytidine oligomers also form four-stranded structures in acidic solution (pH 5) in which interdigitation of the photonated C·CH$^+$ pairs produces the "I-motif" quadruplex. Examples of these two structures are illustrated in Figure elsewhere in this chpater.

1.3 PERSPECTIVES

The double helix was discovered 40 years ago by constructing molecular models consistent with X-ray diffraction patterns obtained from DNA fibers. This view of general conformation remains essentially unchanged, but our increasing appreciation of the complexities of genetic organization and regulation has propelled the search for functionally relevant, local structural features encoded in the nucleotide sequence.

Detailed analysis has been made possible by technological developments both in instrumentation for studying the shapes of large molecules and in the synthetic chemistry and enzymology that make available fragments of DNA of defined sequence.

The intrinsic flexibility of the linear primary structure is potentially a source of great conformational variety that is best explored by analogy with simpler model compounds. Such studies have illustrated a number of sometimes unexpected structural possibilities available to nucleic acids that may be exploited in vivo.

The approach can supply detailed views of fragments of the double helix that contain potentially mutagenic lesions resulting from errant base pairs between unmodified or chemically altered bases. Similarly, complexes of oligonucleotides with drugs and with proteins display details of the molecular recognition events involved.

The ultimate goal is to achieve at the structural level a predictive understanding of the behaviour of DNA in living organisms. Here the molar quantities of material present and the complicating effects of other substances in the environment preclude direct study.

Different model systems and analytical techniques provide alternative perspectives of this problem, supplying complementary information at various levels of detail. Interpretation of the data as a whole offers the possibility that the influences of individual experimental conditions can be identified and that the finer detail supplied by simpler models relate to more elaborate systems that are closer representations of DNA in vivo.

1.4 FORMATION AND STRUCTURE OF NUCLEIC ACID HYBRIDS

Hybridization is the formation of partially or completely doublestranded (duplex) nucleic acid (DNA:DNA, DNA:RNA or RNA:RNA) by sequence-specific interaction of two complementary single-stranded nucleic acids.

"Reassociation" and "renaturation" are terms often used to describe hybridization between completely complementary DNA or RNA strands. Solution hybridization is an integral part of the polymerase chain reaction. Hybridization is often carried out using a labeled probe in an attempt to detect the existence and quantity of complementary sequence in a complex nucleic acid mixture.

The target is often immobilized, as in Southern and other blotting techniques. In this discussion of the physical-chemical aspects of hybridization, in particular thermodynamics and kinetics, melting temperatures, hybridization rates, and dissociation rates and temperatures are interrelated. Details are provided for calculation of these useful properties, as well as examples of calculations.

DNA:DNA, DNA:RNA, and RNA:RNA hybrid duplexes are the products of hybridization reactions between complementary single-stranded molecules. Hybrid formation between completely complementary DNA or RNA strands often is described as reassociation or renaturation. The reverse reaction is called strand separation or dissociation.

The structural differences between deoxyribose and ribose in DNA and RNA strands are reflected in different forms of the antiparallel double helices and in quantitative differences in hybrid stability and the rate of hybrid formation. Nevertheless, they are qualitatively the same, and the basic concepts will be treated together.

1.5 STABILITY OF DNA:DNA, DNA:RNA AND RNA:RNA HYBRIDS

1.5.1 Melting and Dissociation Temperatures

The four *different* temperatures defined here are associated with nucleic acid melting. The corresponding physical phenomena are illustrated in Figure elsewhere in this chapter.

1. t_m^{∞}, the melting temperature of a polynucleotide duplex, such as a long DNA molecule.
2. t_{ss}, the strand separation temperature of a polynucleotide duplex, also called the denaturation temperature in PCR.
3. t_m, the melting temperature at which at least one strand of the duplex is an oligonucleotide.
4. t_d, the dissociation temperature for an oligonucleotide bound to a polynucleotide on a solid support, such as a blot.

1.5.2 Stability of Polynucleotide Duplexes (DNA:DNA, DNA:RNA, and RNA:RNA)

As indicated in Figure elsewhere in this chapter, t_m^{∞}, is the point where the fraction of single-stranded base pairs, f_{ss}, is 0.5, leading to molecules containing alternating duplex and denatured (loops or ends) regions. Because the reaction is intramolecular and equilibrium is achieved, t_m^{∞}, is independent of polynucleotide concentration and time,

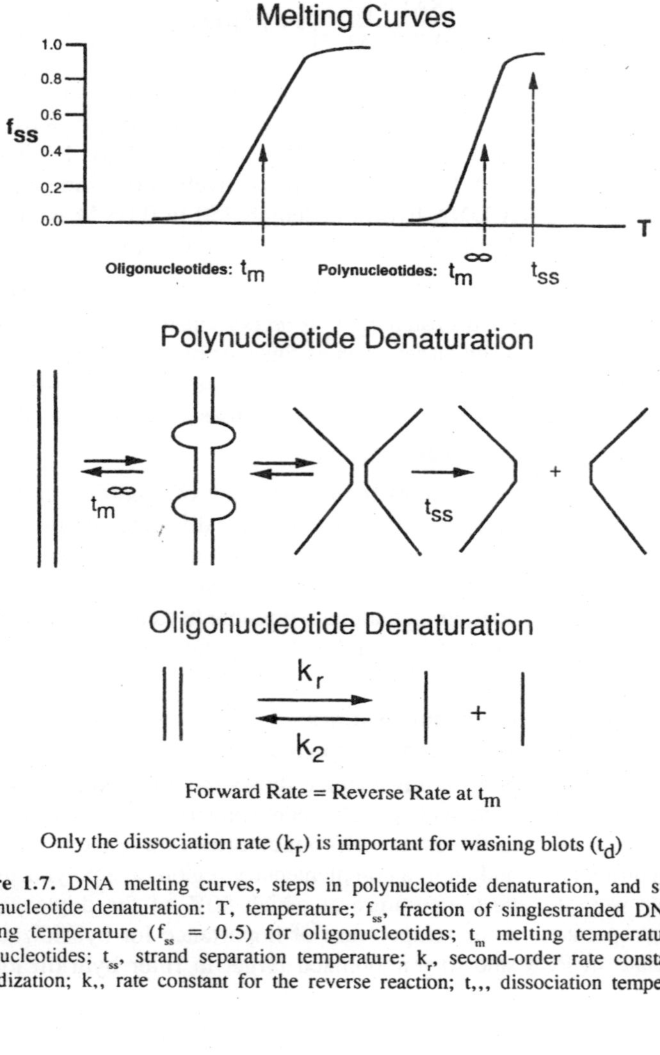

Figure 1.7. DNA melting curves, steps in polynucleotide denaturation, and steps in oligonucleotide denaturation: T, temperature; f_{ss}, fraction of singlestranded DNA; t_m, melting temperature (f_{ss} = 0.5) for oligonucleotides; t_m melting temperature for polynucleotides; t_{ss}, strand separation temperature; k_r, second-order rate constant for hybridization; k,, rate constant for the reverse reaction; t,,, dissociation temperature.

depending only on the composition of the polynucleotide duplex and the composition of the solvent. Strand separation occurs a few degrees above t_m^{∞}, at t_{ss}, and is not a rapidly reversible reaction. For polynucleotides of length *L* containing only G:C and A:T or A:U base pairs, t_m^{∞} (°C) may be determined from the empirical equation:

$$t_m^{\infty} = A^* + 16.6$$

$$\log_{10}\left\{\frac{[\text{SALT}]}{1.0 + 0.7 \cdot [\text{SALT}]}\right\} + B^*(\%G + C) - \frac{500}{L}$$

where [SALT] = $[NA^+] + 41Mg^{2+}]^{0.5}$ and potassium may be substituted for sodium, as in the polymerase chain reaction (PCR); A* is 81.5, 67, or 78°C, and *B** is 0.41, 0.8, or 0.78 °CI(% G+C) for DNA:DNA, DNA:RNA, and RNA:RNA hybrids, respectively. For example, *t;*° for a 50% G+C DNA:DNA hybrid of length 250 in 0.05 M KCI, 0.002 M MgCl, ([SALT] = 0.23), *t* = 81.5 - 11.7 + 0.41(50) - 2 = 88.3°C.

For imperfectly matched hybrids, t_m^{∞} is reduced approximately 1°C for each percent mismatching; t_{ss} is proportionally reduced. Stringency in washing immobilized targets such as filters or blots (Southern, northern, slot) containing polynucleotide probes, a measure of selective dissociation of imperfectly matched hybrids, is increased as the temperature approaches t_m^{∞} in the wash solvent.

Use of different solvents (or introduction of modified bases) may affect A* and/or *B**, with different effects for different polynucleotides. For example, base composition may be eliminated as a variable for DNA:DNA hybrids *(B** = 0) at 3.2 *M* Me4NCI or 2.4 *M* Et4NCI. In another example, R-loop formation is possible because t,,; is decreased by 0.63°C/% formamide for DNA:DNA, but significantly less for DNA:RNA hybrids.

1.5.3 Stability of Oligonucleotides: Oligonucleotide and Polynucleotide

Figure elsewhere in this chapter shows the all-or-none melting of complementary oligonucleotides. The more common experiment developed in this section involves an oligonucleotide, in excess at molar concentration C, binding to a complementary region on a polynucleotide. Either way, t_m, the temperature at which 50% of the duplexes have strandseparated, is the temperature of importance for hybridization to template in PCR and to immobilized target in filter hybridization.

Note that t_m is not t_d, the dissociation temperature important for washing of filters; t_d is discussed in other section of this chapter.

The t_m or hybridization of excess oligonucleotide probe to a polynucleotide target is given by:

$$t_m = \frac{T\Delta H^\circ}{\left(\Delta H^\circ - \Delta G^\circ + RT\ \text{In}[C]\right)} + 16.6$$

$$\cdot \log_{10}\left\{\frac{[\text{SALT}]}{1.0 + 0.7[\text{SALT}]}\right\} - 269.3$$

where enthalpy $\Delta H^\circ = \Sigma_{nn}\left(N_{nn}\Delta H^o_{nn}\right) + \Delta H^o_e$

free energy $\Delta G^o = \Sigma_{nn}\left(N_{nn}\Delta G^o_{nn}\right) + \Delta G^o_e + \Delta G^o_i$

$T = 298.2K$ and $R = 0.00199$ kcal/mol · K

The first term in the t_m equation is T_m (K) in 1 *M* NaCl. The symbols i, nn, and e indicate contributions from initiation of base pair formation, nearest neighbors, and dangling ends. Complete nearestneighbor data sets are available for DNA:DNA and RNA:RNA but not DNA:RNA. Estimated average contributions for each of two possible dangling ends are ΔH^o_e = -5.0 kcal/mol and ΔG^o_e = -1.0 kcal/mol. For the DNA:DNA nearest neighbor values in the following tabulation, ΔG^o_i = +2.2 kcal/mol and

	AA	AT	TA	CA	GT	CT	GA	CG	GC	GG
or	TT			TG	AC	AG	TC			CC
$-\Delta H^o_{nn}$	9.1	8.6	6.0	5.8	6.5	7.8	5.6	11.9	11.1	11.0
$-OG^o_{nn}$	1.55	1.25	0.85	1.15	1.40	1.45	1.15	3.05	2.70	2.30

For example, consider 1 μM $(dG)_6$ binding to $(dC)_6$ in a long DNA at 1 M NaCl. $-\Delta H^o$ = 11.0(5) + 5(2) = 65.0 kcal; $-\Delta G^o$ = 2.3(5) + 1.0(2) - 2.2 = 11.3 kcal; RT In[C] = -8.2 kcal; t_m = 39.9°C.

1.6 FORMATION OF DNA:DNA, DNA:RNA, AND RNA:RNA HYBRIDS

1.6.1 Hybridization with Polynucleotides

The concentration term used for polynucleotide hybridization is molar nucleotide concentration, C_0, not strand concentration, C. The second-order rate constant k_2 ($M^{-1}s^{-1}$), is given by:

$$k_2 = \frac{K'_N\sqrt{L_s}}{N}$$

where L_s is the length of the shortest strand participating in duplex formation, N is the complexity or the total number of base pairs present in nonrepeating sequences, and k'_N is the nucleation rate constant, discussed shortly. The interpretation of N and L is illustrated in Figure elsewhere in this chapter.

The reaction rates are characterized by k_2, whether in solution or immobilized on a solid support such as a Southern blot. When a dilute probe is employed with an immobilized target, stirring may be necessary to maintain homogeneity.

When using a labeled probe and an unlabeled target, the concentration governing the reaction is that of the molecule in excess, the driver. When immobilized DNA is the driver, C_0 is difficult to determine, being a measure of available target concentration. Hybridization reactions are pseudo-first order if the driver is single stranded.

The time course (f_{ss} of the potential duplex target versus time, t) is given as follows:

Second order: Half-time (seconds):

$$\frac{1}{f_{ss}} = k_2\left(\frac{C_0}{2}\right)t + 1 \qquad t_{1/2} = \frac{2}{k_2 C_0}$$

Pseudo-first order: Half-time:

$$f_{ss} = \exp\left[-k_2 (C_0) t\right] \qquad t_{1/2} = \frac{\text{In } 2}{k_2 C_0}$$

For the second-order reaction, a plot of $1/f_{ss}$ versus t gives a slope equal to k_2 $(C_0/2)$. For the first-order reaction, a plot of In(I/f_{ss},) versus t gives a slope equal to k_2C_0.

The rate of hybridization reactions, hence k'_N is zero at t_m^{∞}. The value of k_2 increases to a broad maximum at about 25°C below t_m^{∞}. At lower temperatures, hybridization rates fall off precipitously as a result of decreased availability of nucleation sites. At optimal temperatures, k_N for DNA:RNA hybrid formation is 20 to 50% slower than DNA: DNA hybrid formation, depending on the degree of RNA secondary structure.

RNA:RNA hybridization rates are similarly reduced. Mismatches up to 10% have little effect on hybridization rates. Solvent viscosity and ionic strength also influence k'_N. In solvents with glycerol, ethylene glycol, or formamide, k'_N is reduced because of increased solvent viscosity compared to 1.0 *M* NaCI at 70°C. The dependence of k'_N on salt concentration is given by

$k'_N = \{4.35 \cdot \log_{10}[SALT] + 3.5\}\ 10^5$ for 0.2 - [SALT $\leq$ 4.0

$k'_N = \{4.35 \cdot \log_{10}(1 + 34\ [SALT))\}10^5$ for [SALT] $<$ 0.2

Thus in typical hybridization buffers approximating I M NaCl, $k'_N = 3.5 \ .\ 10^5\ M^{-1}s^{-1}$.

1.6.2 Hybridization with Oligonucleotides

Hybridization with oligonucleotides usually takes place far below t_m^{∞}. (at or below t), so the decrease in k_2 near t_m^{∞} is not a factor, and the temperature dependence of k_2 for oligonucleotides is small unless the

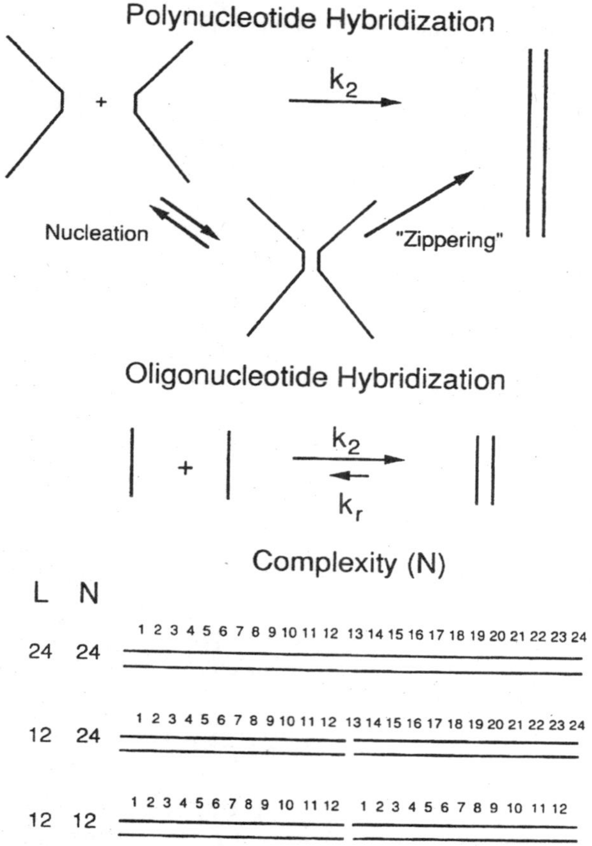

Figure 1.8: Steps in polynucleotide hybridization, steps in oligonucleotide hybridization, and complexity.

target has interfering secondary structure. Oligonucleo tide hybridization has the same salt dependence as polynucleotide hybridization. In fact, oligonucleotide hybridization may be described by the same equations.

For oligonucleotide-polynucleotide reactions, the oligonucleotide is usually the driver leading to a pseudo-first-order reaction characterized by a rate constant with $N = L$ of the oligonucleotide. Oligonucleotide probe reactions are very fast: $k_2 = k'_N/\sqrt{7} = 7 \cdot 10^4\ M^{-1}s^{-1}$ for a 25-mer in 1 M NaCl at t_m^{∞} - 25°C. At 1 nM (C_0 = 25 nM), $t_{1/2}$ = In 2/(7·10⁴·25·10 = 396 seconds (6.6 min). For a 1 μM primer in PCR buffer ($k'_N = 7 \cdot 10^4\ M^{-1}s^{-1}$), $t_{1/2}$ is 3 seconds. For every additional 10°C decrease, k, is reduced 1.2-fold. For I p.M $(dG)_c$ in I M NaCl at t_m, $t_{1/2}$ = 1.3 seconds. Note that $t_{1/2}$ increases proportional to the degeneracy of mixed oligonucleotide probes or primers.

1.7 OLIGONUCLEOTIDE DISSOCIATION

When hybridized oligonucleotides on solid supports are washed to eliminate unhybridized oligonucleotides, both t_m and the rate characterized by k_2 decrease as C decreases, and the hybrid stability depends entirely on the rate constant of dissociation, k_r. In other section of this chaper, we determined the enthalpy of melting, $-\Delta H°$, and t_m which is a measure of the equilibrium constant for oligonucleotide hybridization, or the ratio of the hybridization and dissociation rates.

In other section of this chapter, we determined the k_2, which has a small activation energy. With this information, we may determine k_r. The activation energy for k_r, E_a, is approximately 4.0 kcal/mol plus $-\Delta H°$. The dissociation temperature t_d may be defined as the temperature at which 50% of correctly matched hybrid is released in a specified length of time, t_{wash}.

Because t_d depends on time and not concentration, unlike $t_{1/2}$ for hybridization and t_m t_d is the same whether unique or degenerate probes are being used. It is t_d that may be estimated for extended washing of 14- to 20-mers as 2°C for each dA:dT pair plus 4°C for each dG:dC base pair. The value of t_d may be calculated using:

$$t_d = \frac{T_m}{1 + RT_m \frac{\left[\ln\left(t_{wash}/t_{1/2}\right)\right]}{E_a}} - 273.2$$

For example, for 1 μM $(dG)_c$ in 1 M NaCl, $E_a = -\Delta H° + 4000$ = 69,000 cal/mol, T_m = 313°K, and $t_{1/2}$ = 1.3 seconds. Thus t_d = 19°C for 50% probe released with an hour (3600 s) washing. This calculation depends on all the elements of the physical chemistry of hybridization.

1.8 NUCLEIC ACID SEQUENCING TECHNIQUES

Nucleic acid sequencing, the determination of the order of nucleotides in a molecule of DNA or RNA, may also include the identification of modifications, such as methylation. Nucleic acid sequencing is not usually an end in itself, but it is an important step toward understanding features such as the function, structure, or evolution of the molecule.

For example, determining the sequence of a gene may allow one to predict the amino acid sequence of the protein encoded (although it does not directly provide information on posttranscriptional or posttranslational modifications) and to identify regions of possible secondary structure or sequence motifs that may be recognition sites for proteins involved in control.

Comparison of the sequence with that from other organisms or related genes from the same organism may allow evolutionary relationships to be inferred. It is useful to be able to determine the sequence either of cloned DNA or of molecules isolated directly from a host organism. Sequencing of RNA is also important.

Sequence determination for both DNA and RNA can be done using enzymatic synthesis or cleavage reactions. Most sequencing is of DNA, using enzymatic synthesis. Computer databases of all published (and many unpublished) nucleic acid sequences are maintained by a number of organizations.

1.9 ENZYMATIC SEQUENCING OF DNA

1.9.1 The Principle

A short oligonucleotide, typically 15 to 20 nucleotides, is mixed with the target DNA. To ensure the formation of a duplex by hydrogen-bonded base pairing, the oligonucleotide must be complementary in sequence to the target. Addition of DNA polymerase and 2'-deoxynucleoside triphosphates (dNTPs) allows DNA synthesis by extension of the oligonucleotide (which is referred to as a "primer").

This would simply extend the primer in the 5'-3' direction, were it not for the inclusion of 2',3'-dideoxynucleoside triphosphates (ddNTPs). Although the ddNTPs can be added to a growing DNA chain by means of their 5'-triphosphate, they will terminate the synthesis of that chain because they carry an-H on the 3' carbon rather than the-OH, which is needed for addition of further nucleotides. Four separate primer extension

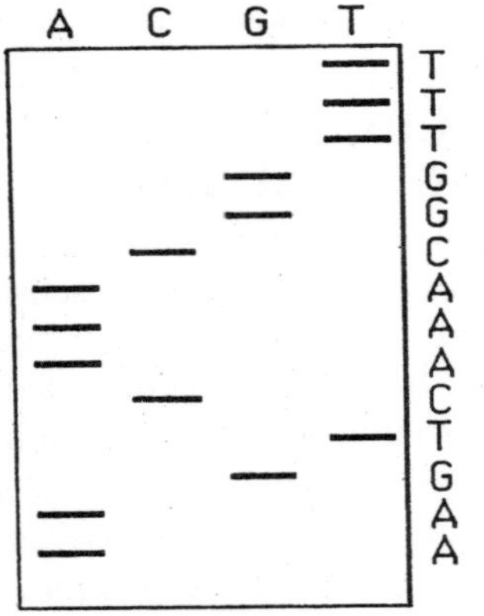

Figure 1.9: Schematic representation of part of an autoradiograph of a sequencing gel: the sequence is indicated up the side.

reactions are done, each containing a different ddNTP: ddATP, ddTTP, ddGTP and ddTTP. Each reaction will generate a "nested" series of newly synthesized molecules, each beginning with the primer and ending at an A (opposite a T) in the tube containing ddATP, at a C (opposite G) in the tube containing ddCTP, and so on.

Note that termination does not necessarily take place opposite the first T in the reaction containing ddATP, because dATP is also present. Some of the extending molecules will incorporate the deoxy form and will progress to the next T, where termination likewise may or may not take place.

The size distribution of the molecules produced will depend on the ddNTP/dNTP ratio: the higher the ratio, the lower the average length of molecules generated. Following the synthesis reactions, the products are separated electrophoretically according to size in adjacent lanes of a polyacrylamide gel.

To prevent the formation of secondary structures in the DNA, which would cause anomalies in the mobility, the electrophoresis is carried out under denaturing conditions (at high temperature and in the presence of urea). Thin gels are used, allowing greater resolution, and molecules differing in length by a single nucleotide can be separated. The positions of the molecules in the gel are then detected.

Usually one of the dNTPs used in the reactions is radioactively labeled (other possibilities for labeling are radioactive end labeling of the primer, and fluorescence labeling), and the gel is autoradiographed. A schematic representation of a result is shown in Figure elsewhere int his chapter. Reading up from the bottom allows the sequence to be deduced one nucleotide at a time.

1.9.2 Generation of Template

It is often convenient if the template for sequencing can be obtained in single-stranded form, and a number of vectors are available for this, based on the filamentous bacteriophage M 13. This phage attacks only "male" *E. coli* cells, adsorbing to the sex pilus. The infectious particles contain single-stranded DNA, but upon entering the bacterial cell this is converted to a double-stranded replicative form (RF).

This material is used to synthesize either more RF DNA or single-stranded DNA, which can then be packaged into phage heads. These are extruded from the bacterial cell without lysing it. In the lab, DNA is cloned into the RF molecule and introduced artificially into a suitable *E. coli* host, after which the phage particles containing ssDNA are collected.

The DNA is then purified and used in the sequencing reactions. The phage vectors contain a "multiple cloning site" with recognition sequences for a range of restriction enzymes allowing convenient insertion of blunt- or sticky-ended fragments. This multiple cloning site is located within a part of β-galactosidase gene *(laeZ)* engineered into the phage.

This "minigene" produces a polypeptide that can assemble with a fragment encoded by the host to produce functional β-galactosidase enzyme. Insertion of DNA into the multiple cloning site therefore abolishes the production of β-galactosidase in infected hosts (which are unable to produce the complete enzyme themselves), and this result can be detected on indicator plates.

Other vectors are available that are hybrids between plasmids and phage. They are propagated as plasmids (and contain a similar multiple cloning site to the M13 vectors). However, they also contain an origin for single-stranded DNA synthesis derived from a filamentous phage. When cells carrying these plasmids are superinfected with a helper phage, this origin is activated and singlestranded DNA is synthesized.

Single-stranded DNA can also be obtained by polymerase chain reaction (PCR) techniques. A convenient method is to have an excess of one of the primers, resulting in preferential amplification of the corresponding strand. Sequencing of double-stranded DNA is also possible. The two strands are first separated by heating or treatment with alkali, and the primer allowed to anneal. Since it will anneal to only one strand, sequencing can proceed, provided sufficient primer-template complex can be formed without reannealing of the full-length strands.

1.9.3 Primers

It is clear that to construct a primer that will anneal to the target molecule, some sequence information is necessary at the start. For material cloned into a standard vector, this is not a problem, inasmuch as the sequence flanking the cloning site will be independent of the insert.

A single artificially synthesized "universal primer" that anneals close to the cloning site can therefore be used. Alternatively, if the sequence of part of the target DNA is already known, a primer complementary to that can be synthesized.

1.9.4 Sequencing Strategies

The amount of information generated in any one sequencing run is limited to a few hundred nucleotides. Most sequencing projects require the determination of larger molecules than this, and the overall sequence is built up in small pieces. This can be done in any of the following ways (and often a combination is used):

1. The larger molecule is fragmented by sonication or restriction enzyme degradation, and the pieces cloned and sequenced.
2. The larger molecule is systematically shortened by exonucleolytic degradation.
3. The larger molecule is cloned intact, and a series of internal primers is used.

In general, when determining any new piece of sequence it is desirable to ensure that both DNA strands have been properly sequenced and that the individual readings for one strand form a continuous overlapping series. Errors caused by secondary structure in the template may not be detected on one strand but will become apparent when it is compared with the other.

1.9.5 Enzymes and Nucleotides

Initially, sequencing was carried out using the Klenow fragment of DNA polymerase. This enzyme had the disadvantage of giving bands of rather uneven intensity, particularly in certain sequence contexts. Subsequently, other enzymes were used which gave more even hand intensities, and perhaps the most popular is the polymerasc encoded by bacteriophage T7, marketed commercially under the name Sequenase.

The use of thermostable polymerases allows the sequencing reactions to be carried out at higher temperatures, reducing the likelihood of artifactual termination due to secondary structure in the template. This procedure has also led to the development of "cycle" sequencing: repeated rounds of synthesis (in the presence of ddNTPs) carried out by cycles of heating and cooling, resembling NCR (although amplification is linear rather than geometric, inasmuch as only a single strand is synthesized in each cycle).

Cycle sequencing is particularly useful when only small amounts of template are available. Nonstandard nucleotides, such as dITP, can be used in all cases to reduce problems caused by secondary structure.

1.9.6 Data Collection and Analysis

There are three distinct requirements. The first is the acquisition of data from individual electrophoresis runs. The second is the assembly of these datasets to produce the overall sequence of the DNA under consideration, and the third is the analysis of this sequence (e.g., to deduce coding function).

Acquisition can be done fully manually, by simply reading the gel from the bottom up, according to whether successive bands occur in the A, C, G, or T track, and entering the information into a computer file either through a keyboard or a modified mouse. This method has minimum equipment requirements but may be subject to user error (depending on the user!).

Acquisition can be done in a semiautomated way, using a sonic digitizer. In this device, a digitizing pen is pressed onto each band in turn. The signal emitted upon pressing is picked up by two detectors. This allows the calculation of the position of the pen, and therefore the lane it was indicating.

The information thus acquired is fed directly into a computer file, allowing the reader to exercise judgment in reading the gel (assessing the significance of possible artifact bands, etc.) while reducing data transcription errors. Acquisition can also be fully automated, by electronic scanning. A large number of computer packages are available for assembly of individual gel readings into an overall sequence and its subsequent analysis.

1.9.7 Automation

The manipulations involved in DNA sequencing are essentially independent of the particular sequence to be determined, and the procedure is therefore particularly suitable for automation. This is important for larger scale projects, such as the sequencing of complete genomes. Automation is currently least widespread in template preparation. Procedures being developed include the use of magnetic particle labeling of DNA for purification.

Degrees of "automation" of the actual reactions vary from the use of predispensed reagents in microtiter trays and hand-operated multiple pipetters to robot pipetters. Gel reading is quite widely automated. Many systems rely on fluorescent labeling of either the primers or the ddNTPs themselves. A different label is used for each reaction, giving each set of products a different fluorescence pattern.

The products of each reaction are mixed (if the fluorescence tags are on the ddNTPs, all four reactions can be done in the same tube) and electrophoresed in a single gel lane. Fluorescence detectors pick up individual bands as they pass through the gel, and determine which nucleotide was present.

The information is recorded directly onto computer disk and displayed visually in a series of traces of different colours. A black and white example is shown in Figure elsewhere in this chapter. By allowing data capture in real time, this approach removes the need for autoradiography.

1.10 CHEMICAL SEQUENCING OF DNA

Sequencing is possible by base-specific cleavage, an approach developed by Maxam and Gilbert in the late 1970s. A DNA molecule is labeled at one end and treated with limiting amounts of reagents, which react specifically or preferentially as follows:

G	dimethyl sulfate
A + G	piperidine formate
C + T	hydrazine
C	hydrazine and 1.5-2 M NaCl
A + C (stronger at A)	1.2 M NaOH at 90°C

Subsequent treatments bring about cleavage at the modified bases, generating nested sets of molecules as with enzymatic sequencing. They are likewise electrophoresed and their positions determined by autoradiography.

Because this technique does not require cloning of the target DNA, it is particularly useful for the sequencing of artificially synthesized oligonucleotides. It is also useful for studies of secondary structure and DNA-protein interactions, both of which will influence the reactivity of individual nucleotide positions.

It can be used to probe modifications such as methylation in the same way. Analysis of genomic DNA is possible by this method. For example, the total genomic DNA is digested with a suitable restriction enzyme and then subjected to cleavage reactions. The fragments are with a labeled probe for the region to be sequenced. The fragments picked out will have one end in common (generated by the restriction enzyme), and the other will be sequence specific.

1.11 RNA SEQUENCING

Like DNA, RNA can be sequenced by enzymatic synthesis or degradative methods. In the former, a suitable primer (which therefore requires some preliminary sequence information to be available) is annealed at the 5' end of the RNA, and sequencing reactions are carried out as with DNA sequencing except that reverse transcriptase is used instead of DNA polymerase.

The products of the four reactions are analyzed electrophoretically exactly as before. Degradative methods employ a range of chemical and enzymatic cleavage reagents followed by electrophoretic or chromatographic analysis.

2

DNA Replication and Transcription

Since an organism's genetic information is stored in its DNA as a nucleotide sequence array, the cell must replicate its chromosomal DNA very precisely. The self-complementary nature of DNA is the basis for accurate replication. DNA polymerase adds deoxyribonucleotides to preexisting DNA chains in the 5'-3' direction.

One strand is synthesized continuously (the leading strand), while the other is synthesized discontinuously (the lagging strand). To extract information from the DNA, a limited region of DNA, corresponding to a gene or gene cluster, is transcribed into RNA.

This process is catalyzed by RNA polymerase, a multisubunit enzyme that adds ribonucleotides to the growing RNA chain in the 5'-3' direction, along the template DNA. Both starting point and the frequency of transcription are determined by the promoter, a specific sequence of DNA to which RNA polymerase and transcription factors bind.

2.1 DNA REPLICATION

Replication of DNA is a key event in cell propagation. A matching set of chromosomal DNA must be transferred precisely into each of the daughter cells. This is achieved only when chromosomal DNA is accurately replicated, providing two copies of the entire genome, with one copy faithfully distributed into each daughter cell.

To this end, the cell has special mechanisms to keep the fidelity of DNA replication very high, to segregate the replicated DNA, and to tightly coordinate DNA replication and cell division within the cell cycle.

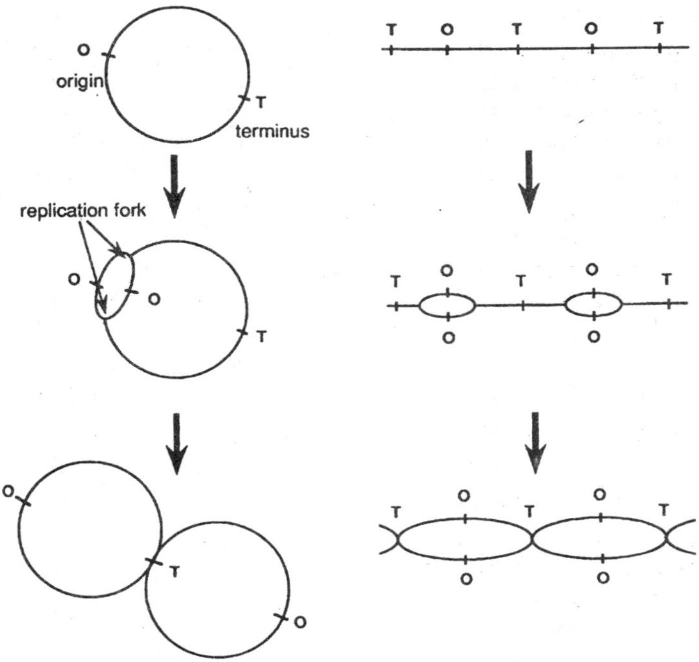

Figure 2.1: Units of DNA replication in prokaryotic and eukaryotic cells: O, replication origin: T, replication terminus.

2.1.1 Structural Aspects of DNA Replication

DNA replication starts at a particular site on the DNA molecule called the origin of replication. In prokaryotic cells (e.g., bacteria), the origin is located at a unique site on the circular chromosome, and DNA replication proceeds bidirectionally from the origin to a terminus site.

On the other hand, eukaryotic cells have many replication origin sites on each chromosome. In both prokaryotes and eukaryotes DNA synthesis is catalyzed only at the replication fork. In duplex DNA, the replication fork advances in a semidiscontinuous manner.

One strand, the leading strand, is synthesized continuously and the other, the lagging strand, is synthesized discontinuously.

2.1.2 Biochemistry of DNA Replication

Many proteins are involved in DNA synthesis at the replication fork. In *Escherichia coli,* more than 20 different proteins participate in this process. Among them, DNA polymerase III holoenzyme, which consists of 10 distinctive polypeptides, is responsible for elongation of the DNA chain in both leading and lagging strand replication.

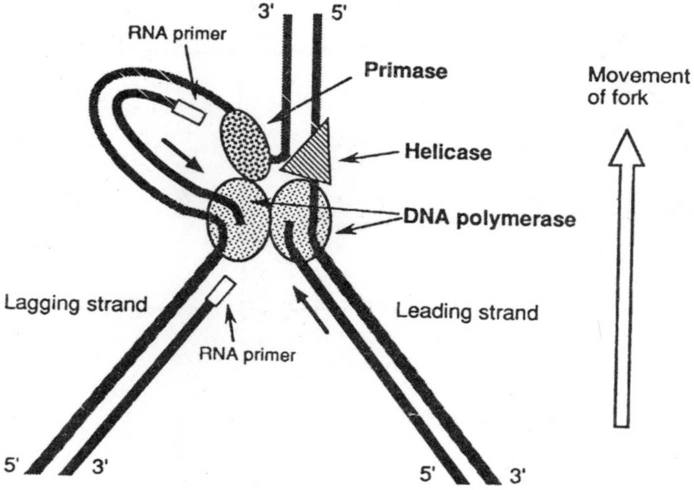

Figure 2.2: Simplified structure of the replication apparatus of Escherichia coli.

This enzyme extends the DNA chain with a high processivity and a high catalytic efficiency, in contrast with the other *E. coli* polymerases, DNA polymerases I and II. A 3'-5' exonuclease activity is associated with polymerase III and provides the enzyme with a proofreading capacity to correct replicational errors.

DNA polymerases 6 and e, replicative enzymes recently found in eukaryotic cells, possess similar characteristics. The DNA polymerase requires a single-stranded region of DNA as a template, and a piece of RNA or DNA annealed to the template as a primer. For this reason, two other enzymes are essential to double-stranded DNA replication.

DNA helicase opens the duplex DNA to provide a single-stranded region at the replication fork, while primase synthesizes a short RNA transcript that acts as a primer for DNA chain elongation. Since chain elongation by DNA polymerase occurs only in the 5'→3' direction, lagging-strand DNA synthesis proceeds opposite to replication fork movement.

This property requires a repeated synthesis of primer RNA for initiating synthesis of the nascent short fragment (Okazaki fragment) on the lagging strand. A concurrent DNA synthesis of both leading and lagging strands is achieved by the formation of a complex called the replisome.

In addition to the components of the replisome, several auxiliary proteins are needed for DNA replication. Single-stranded DNAbinding protein (SSB) stabilizes the structure of the replication fork, while DNA topoisomerase, acting as a swivel, releases torsional stress produced ahead of the replication fork.

To complete the discontinuous DNA synthesis, RNase H (for removing primer RNA), DNA polymerase I (for filling the gap), and DNA ligase (for connecting the nascent fragments) are required.

2.1.3 Regulation of DNA Replication

DNA replication is principally regulated at the step of initiation of synthesis. The initial signal that turns on DNA replication involves growth factors in eukaryotic cells, or nutrition in prokaryotic cells, but it is largely unknown how these signals are transduced to the initiation event.

As revealed with an *E. coli* reconstituted system, at least eight proteins are required for replication of plasmid DNA carrying on C. the unique 245 by chromosomal origin of *E. coli*. A key event in initiation is formation of a complex in which the initiator protein (DnaA protein) tightly binds to the 9 by dnaA boxes in the *on C* sequence.

Following this step, duplex DNA is opened in an ATPdependent manner, and a helicase and SSB are settled in the singlestranded region to form a prepriming complex. Primase and DNA polymerase III holoenzyme then start DNA synthesis. A similar mechanism has been found in the initiation of SV40 (a eukaryotic virus) DNA replication in vitro; large T antigen acts as a helicase as well as an initiator protein specifically bound to the SV40 origin.

Although DNA replication origins and proteins binding to them have been found in yeast *Saccharomvces cerevisiae,* the initiation mechanisms in eukaryotes are largely unknown.

2.2 TRANSCRIPTION

Transcription is the process of synthesizing an RNA molecule complementary to its template DNA, whereby the transfer of information from DNA to protein begins. DNA-dependent RNA polymerase binds to a specific DNA sequence called the promoter; then it unwinds the duplex DNA for about one turn of the helix, to expose a short stretch of single-stranded DNA so that complementary base pairing can be made with incoming ribonucleotides.

The enzyme joins two of the ribonucleoside triphosphate monomers and then moves along the DNA strand, extending the growing RNA chain in the 5'-s3' direction until it encounters a second special sequence, called the terminator, which signals where RNA synthesis should stop.

Table 2.1: Subunit composition of the *E. coli* RNA polymerase.

Subunit	*Gene*	*Number of Amino Acids*	*Molecular Weight*	*Function*
α	rpo A	329	36,512	Connecting ββ′-subunits
β	rpo B	1342	150,618	Catalyzing RNA synthesis
β′	rpo C	1407	155.163	Template binding and association with σ-subunit
σ^{70}	rpo D	613	70,263	Recognition of general promoters

After transcription has been completed, each RNA chain is released from the DNA template as a free, singlestranded RNA molecule.

2.2.1 RNA Polymerase

Prokaryotic cells are equipped with a single type of RNA polymerase for transcription of genes, beside DnaG primase for synthesis of RNA primer required for initiation of DNA replication. The RNA polymerase of *E. coli* is composed of multiple subunits, with a total mass of about 450 kDa.

The holoenzyme, the catalytically active form with promoter selectivity, consists of five subunits: two cs-subunits, and one each of (β, β′, and σ. The rrsubunit, which ensures that the holoenzyme binds to the proper site in the promoter, is released once eight or nine nucleotides have been joined. There are multiple forms of sigma, each responsible for recognizing a particular class of promoters.

The predominant form, serving for general promoters, is σ^{70}. In eukaryotic cells, transcription is accomplished by three distinct types of RNA polymerase. RNA polymerase I (pol 1) makes ribosomal RNA (18 and 28S rRNA) and pol II mostly makes messenger RNA, while pol III makes transfer RNA, 5S ribosomal RNA, and a number of other small RNA molecules.

All these enzymes are large and complex, with molecular weights of about half a million. For example, RNA polymerase II contains two large subunits (~200 and ~140 kDa) and a collection of smaller polypeptides, each less than 50 kDa. The largest subunit shares amino acid sequence homology with the β-subunit of the *E. coli* enzyme, while the second largest is related to β′.

No eukaryotic RNA polymerase can initiate transcription by itself, and the initiation at proper sites requires a group of general initiation factors. For pol II, at least six general transcription factors (TFIIA, TFIIB, TFIID, TFI[E, TFIIF, and TFII-I) are required. The polymerase enzyme together with the general factors constitute a basic transcription apparatus that can initiate transcription at the proper sites.

2.2.2 Transcription Units

Transcription takes place within a limited region of the genomic DNA, and generally only one of the two DNA strands is used as a template. The promoter determines the starting point of transcription as well as which of the two strands is copied.

Table 2.2: Fukaryotic RNA Polymerases.

Polyin-crase	*Location*	*Copies per Cell*	*Products*	*Polymcrase Activity of Cells (%)*
pol I	Nucleolus	~40,000	35S and 47S pre-ribosomal RNA	50-70
pol II	Nucleoplasm	~40,000	Heterogeneous RNA (premessenger RNA) Small nuclear RNA (Ul, U2, U4, U5)	20-40
pol III	Nucleoplasm	~20,000	Transfer RNA, 5S RNA, 7S, RNA, snRNA (U6), other small RNA Molecules	~10

A transcription unit extends from the promoter to the terminator, where the RNA polymerase stops adding nucleotides to the growing RNA chain. A critical feature of the transcription unit is that it constitutes a stretch of DNA transcribed as a single RNA molecule, as shown in Figure elsewhere in this chapter. A transcription unit may include only one gene or several.

Bacterial promoters are identified by two short conserved sequences, centered at -35 and -10 nucleotide relative to the startpoint. Promoters whose -35 and - 10 regions closely approximate the consensus sequences (TTGACA for -35 and TATAAT for -10 in *E. coli*) are strong promoters, whereas promoters whose sequences differ significantly from the consensus sequences are weak.

Bacterial RNA polymerase terminates transcription at sites of two types. Factor-independent sites contain a GC-rich hairpin followed by a run of U residues, while factor-dependent sites require rho factor.

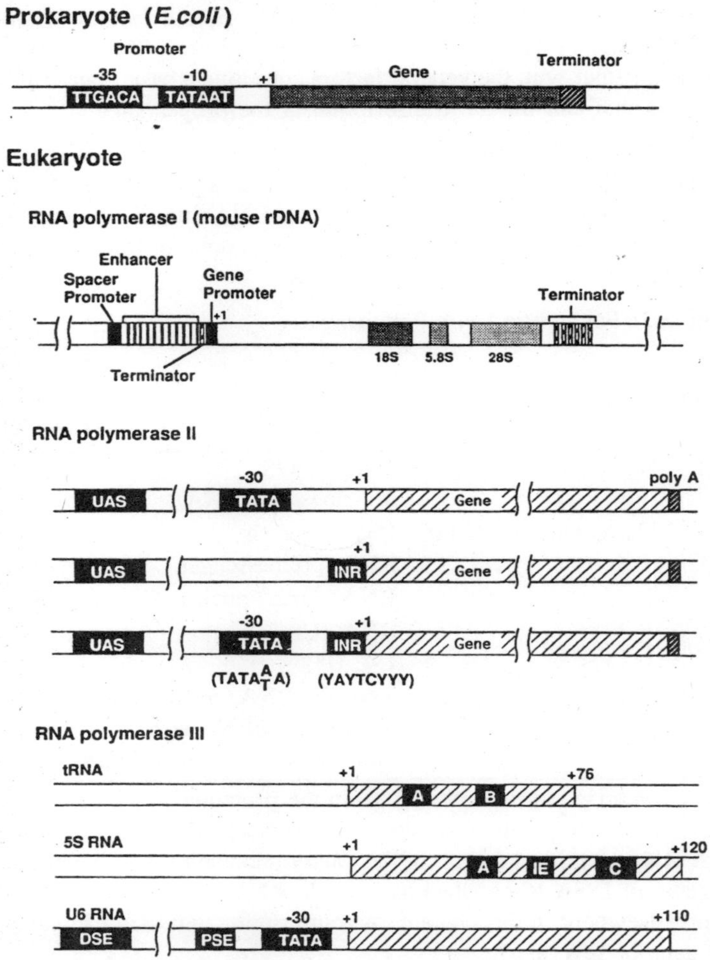

Figure 2.3: Transcriptional units in prokaryotic and eukaryotic cells: LAS, upstream activating sequence; TATA, TATA box; INR, initiator; A. B, C, and IE. intragenic promoter elements for RNA polymerase Ill; DSE, distal sequence element, PSE, proximal sequence element; Y, pyrimidine.

In eukaryotes, the promoters recognized by each RNA polymerase are distinct. The promoters for RNA polymerases I and II are mostly upstream of the starting point, whereas the promoter for RNA polymerase III most often lies downstream. Promoter elements that are necessary and sufficient for specific initiation by RNA polymerase II with its general factors are referred to as minimal, or core, promoter elements.

The most common core elements are the TATA box found about 30 nucleotides upstream of the transcription start site and less well-characterized initiator (INR) elements at the start site.

Specific genes may contain either or both of these elements, which may function together with other cis elements. Little is known about the termination process for eukaryotic RNA polymerases, even their termination signals.

2.2.3 Regulation of Transcription

Initiation of RNA synthesis is a primary control point in the regulation of differential gene expression. Cells respond to intra- and extracellular signals by turning certain genes on or off and modulating the extent of transcription of active genes. In bacterial cells, there are several different mechanisms by which modulation of transcription initiation is achieved: exchange of sigma factors, inactivation of repressor protein, activation of transactivator, and conversion of the DNA topology by topoisomerase.

In eukaryotic cells, an assortment of regulatory elements for RNA polymerase II transcription can be scattered both upstream and downstream of the transcription start site for a gene. Each gene has a particular combination of positive and negative regulatory cis elements that are uniquely arranged with respect to number, type, and spatial array.

These elements are binding sites for sequencespecific transcription factors that activate or repress transcription of the gene. The DNA binding proteins also interact with activator proteins that also interact with the core transcription apparatus (e.g., TAFs).

Usually, cis elements are arrayed within several hundred base pairs of the initiation site, but some elements can exert their control over much greater distances (1-30 kb). The control region in the vicinity of a transcription start site is called the promoter as described earlier, and regions that regulate the promoter activity from a distance and in an orientation-independent fashion are called enhancers. In addition, transcription can be regulated by the premature termination of elongation, or attenuation.

2.3 DENATURATRION OF DNA

The *denaturation of DNA* is brought about by any of the many artificial means of disturbing the noncovalent interactions supporting the native conformation, causing either alterations at the level of local structure or the total collapse of the double helix. The process is generally reversible.

2.3.1 Denaturation and Degradation

The native helical conformation of DNA is supported by noncovalent forces that are substantially weaker than the covalent forces that hold the polynucleotide chains together. For the most part, this noncovalent support consists of a short-range dispersion forces sensitive to temperature.

Watson-Crick hydrogen bonds between bases maintain the alignment of the two strands, but the vertical ''stacking'' forces between nearest-neighbor base pairs are the major sources of support to the helix. Stacking forces vary with the sequence of stacked pairs in the duplex.

There are 10 unique nearest-stacked-neighbors, each with a different stacking energy. These differences lead to variations in local stability of the DNA as well as a strong dependence of the structural state of each pair on the state of its neighbors. Indeed, the dependence extends well beyond next neighbors.

The energy needed to free a pair from the stacked array is so great that when provided, it leads to the cooperative collapse of entire stretches that may involve as many as 250 to 500 contiguous pairs. Factors used to break these bonds usually bring it about over a very narrow range of conditions and affect large stretches of the DNA.

In addition to stacking and hydrogen bonds, a number of weaker bonds contribute to the support of the duplex, primarily From .solvation effects, the occasional water bridge, and hydrogen bonds associated with certain sequence arrangements. Also, the phosphodiester groups (...—O—P[O_2]—O—...) that join nucleotides in the backbone have a negative charge; therefore the polynucleotide chain is a polyanion with strong repulsive forces between opposing chains.

Counteracting the stabilizing forces from stacking and hydrogen bonds. The high charge density of the helix causes counterions (e.g., Na^+, K^+, Mg^{2+}) to condense on the helix in high concentration, effectively reducing the net charge on the phosphates and screening the repulsion between negative charges.

If these weak stabilizing forces are broken artificially. the double helix collapses and the DNA is denatured. Denaturation leads to nonhelical structures and, if sufficiently severe, to total separation of the polynucleotide chains. If till noncovalent bonds are broken, the chains assume the conformation of random coils. However, denaturation is rarely if ever so complete; rather, the DNA assumes a complex disordered state where residual noncovalent bonds are mostly misdirected.

If DNA is forced into an altered structure, one that is not everywhere helical or everywhere recognizable by the various enzymes and proteins

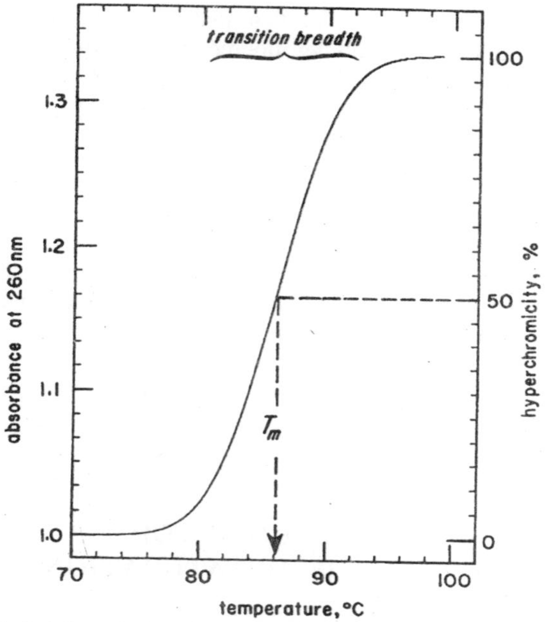

Figure 2.4: Typical thermal denaturation (melting) curve for DNA, showing the variation in optical absorbance at 260 non with increasing temperature. The T , usually defined as the temperature corresponding to half the overall change in absorbance, has physical significance only for heterogeneous DNAs of considerable size.

that maintain or interact specifically with the native structure, it is still said to be denatured. Noncovalent bonds can be broken through one of four means, involving exposure of the DNA to high temperatures, extreme pH values, low counterion concentrations. and various denaturing solvents.

Denaturation is distinguished from the more severe process of degradation, during which covalent bonds are broken. Thus to avoid degradation, denaturation of DNA by high temperatures for studies requiring the integrity of the sequence he maintained (e.g., heteroduplex formation, sequencing, and the polymerase chain reaction) should involve a minimum exposure of the single strands to temperatures much above 50°C.

Single-stranded DNA is particularly prone to depurination at high temperature and low pH. Many procedures for denaturing DNA call for heating in boiling water for 2 to more than 10 minutes. The incubation time need only ensure that thermal equilibrium has been achieved, since denaturation is almost instantaneous. The structure of native single-stranded DNA is more complex than that of duplex DNA, with many regions of single-strand stacking, double-strand hairpins, and segments of random and dynamic conformation.

Single-stranded DNA adopts these local structural features according to the sequence. Double-stranded DNA collapses into a similar but metastable, quasi-stochastic structure if quick-cooled from high temperature. Complete single-strand stacking does not occur until well below room temperature.

Also, since the conformational states of stacked bases in single-stranded DNA are essentially independent of those of their neighbors denaturation occurs over a wide range of conditions. Melting is gradual, occurring over a range of 80 to 100°C, whereas duplex DNA typically melts within 12 to 16°.

Indeed, if the sequence is sufficiently uniform, duplex DNA may melt within 0.1°C. Most physical and chemical properties of single-stranded DNA are like those of RNA.

2.4 ANALYTICAL METHODS FOR FOLLOWING THE DENATURATION PROCESS

Methods for monitoring the denaturation of DNA can be classified as short-range methods (those that depend on specific short-range physical properties), long-range methods (which reflect alterations in the macromolecular properties), chemical methods (which probe structural alteration by changes in reactivity) and biological methods (which rely on sensitivities to various biological probes such as nuclease sensitivity).

Differential scanning calorimetric and various spectroscopic methods probe short-range effects, while hydrodynamic and electrophoretic methods monitor the long-range effects. The most common method of monitoring the denaturation process is by *absorbance spectroscopv* in the ultraviolet region of 200 to 290 urn.

The mean absorption coefficient for $\pi \rightarrow \pi^*$ electronic transitions of the heterocyclic bases is unusually large over this region. In the helical state these transitions are neighbor-perturbed in the close-ordered packing of the bases. The molar absorption coefficient for the helix, $\varepsilon_{260nm}(P)$, is approximately 7 × 10^3L/ mol·cm under physiological solvent conditions.

During denaturation, the bases unstack and the molar absorption coefficient increases to ~ 10^4L/mol·cm, close to that for the free nucleotide monomers. The increase at 260 nm is approximately 38%, which means that denaturation can be monitored with considerable precision on nucleotide residue concentrations in the micromolar range.

A typical thermal denaturation or melting curve of DNA is shown in Figure elsewhere in this chapter. A melting temperature T_m, determined from the midpoint of the transition, is generally taken as a measure of the thermal stability of the DNA; while the transition breadth is taken as a measure of the variation in (G+C) content.

Fine detail is evident in the high resolution, first-derivative melting profile of Figure elsewhere in this chapter, which shows the melting of pN/MCS-6, a recombinant plasmid of 4750 base pairs. Denaturation takes place in saltatory fashion, by domains of increasing stability with increasing temperature.

A domain in this instance is a particular stretch of DNA that dissociates or melts in a cooperative, all-or-none fashion; it is seen in profiles such as this one as sharp peaks representing subtransitions. Methods have been devised for assigning subtransitions to specific sequences. The Tdetermined from the midpoint of integration in Figure elsewhere in this chapter is now seen to have no particular significance for the stability of specific sequences.

The subtransition for denaturation of a 330 by (A·T)-rich insert can be seen clearly at 76.40°C. The temperatures at which different domains melt depend on a number of factors, including the frequencies of the 10 unique nearest neighbors and the number of changes in helix-coil boundaries separating domains. The insert domain of Figure elsewhere in this chapter melts as an internal loop, generating two new boundaries.

The T_m for the insert subtransition is only 75.11°C when the insert melts without generating new boundaries (i.e., when it is located at the end of the DNA). *Circular dichroism,* also effective in the region of 200 to 290 run, measures the difference in molar extinction of the left and right circularly polarized components of monochromatic light.

As such it is used to detect the spatial asymmetry and helical handedness as the chromophoric base pairs twist one way or the other around the major axis. Data from in and *Rainan spectroscopy* in the 500 to 1800 cm^{-1} region reflect changes in *vibrational* and *rotational* modes affected by changes in hydrogen bonding and stacking interactions.

Changes involving polar groups give the most intense absorptions and are useful for following the dissociation of hydrogen bonding. There is a superabundance of spectral detail in these spectra, since the number of transitions for all possible modes approaches 3n – 6, where *n* is the number of atoms. The downside is that extinction coefficients are typically

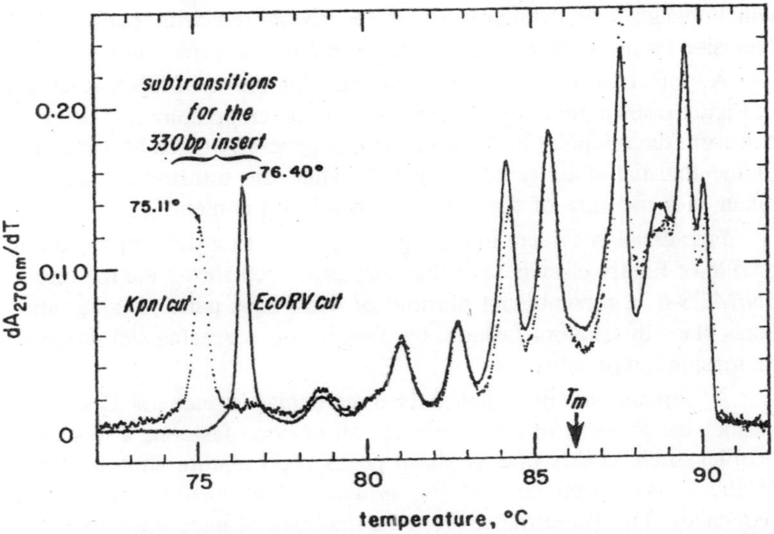

Figure 2.5. The example of a melting curve of a homogeneous DNA of discrete length (4750 bp), obtained and plotted as the derivative of the change in absorbance with temperature, $dA_{270\ nm}/dT$. At 270 nm the change in absorbance for dissociation of A·T base pairs is equal to the change for G·C pairs.

several orders of magnitude smaller than in the electronic region, with the result that DNA concentrations need to be higher.

Working concentrations are usually 0.01 to 1.0 mg of DNA per milliliter, depending on the transition and sensitivity of the facility. *Nuclear magnetic resonance,* the measure of circular motion of magnetic nuclei due to their interaction with the magnetic component of electromagnetic radiation in the radio frequency range (10^{10} Hz), is the most powerful method for investigating dynamic structural changes during denaturation.

One-dimensional spectra in the range of 9.5 to 0.5 ppm are used mainly for studies of the exchangeable imino and amino resonances, important indicators of the hydrogen-bonded states of the bases. Phase-sensitive, twodimensional NMR in D,O is used for tracing single-step connectivities and for measuring coupling constants.

Since there are only the four bases in DNA, it is difficult to assign spectral features to specific bases in a sequence of any length. To reduce or eliminate ambiguity in assignments, specimens usually are oligomeric, and one must contend with end effects.

Hydrodynamic methods reflect the reduction in frictional coefficient during denaturation when DNA changes from a rigid rod into flexible chains. *Viscometry, sedimentation,* and *dynamic light scattering* are generally more difficult and cumbersome than spectral tral methods, requiring greater amounts of monodisperse material.

The long-range effects of denaturation also have been studied visually by *electron microscopy*. The methodology involves fixing partially denatured regions at intermediate temperatures with glyoxal. This bifunctional reagent reacts with imino and amino groups, preventing re-formation of the helix.

Denaturing gradient gel electrophoresis (DGGE), a new technique that also monitors changes in conformation during denaturation, has been used to detect minor sequence variations and mutations. The sensitivity of this technique to small thermodynamic differences between specimens is approximately the same as that obtainable by electronic absorption spectroscopy.

However, in this instance the sensitivity is most effective in detecting small differences between specimens. In parallel DGGE a gradient of denaturing conditions is established between the electrodes at the two ends of a (vertical) slab gel. The gradient is approximately 30 to 70% (v/v) formamide plus 7 M urea, and the run is carried out at 60°C.

During electrophoresis, the mobility of the specimen DNA remains constant until one of the domains denatures. When the DNA enters a region of the gradient that leads to denaturation of the weakest domain, the mobility of the DNA abruptly decreases. Two fragments with a limited number of distinct melting domains, differing by a single base pair, have been clearly resolved by DGGE.

2.5 PHYSICAL AND CHEMICAL FACTORS AFFECTING THE STABILITY OF DNA

A number of factors from within and without the DNA affect its sensitivity to denaturation:

1. The (G+C) content, or more precisely the distribution of the 10 nearest-neighbor frequencies in the DNA.
2. The local sequence and conformational state of neighboring domains.
3. The presence of wobble or other mispairs.
4. The DNA length.
5. The counterion or cation type and concentration.
6. pH.
7. Temperature.

8. Various salts and nonaqueous solvents.
9. Supercoiling stresses.

The first denaturation experiments were performed by titration with acid and alkali. Protonation and deprotonation of the bases at the upper and lower extremes of hydrogen ion concentration alter the hydrogen bond donor-acceptor relationships between pairs and place charges on the bases. DNA denatures below about pH 4 and above about pH 11 somewhat higher and lower, respectively, than the *pK* values of the free bases.

The difference between *pKs* for the free bases and pHfor denaturation reflects the free energy of isolating the ionizable groups by Watson-Crick hydrogen bonding. There are practical reasons for avoiding pH values below about 5, however: protonation leads to precipitation of intact polymeric DNA, while still lower pHs increase the rate of depurination and subsequent chain cleavage. Denaturation at high pH, on the other hand, appears to be less problematic and represents a method for separating the helical strands.

Provided the strands differ from unity in the ratio of purines to pyrimidines, they can be separated by CsCI buoyant density gradient centrifugation at about pH 11, where thymine and guanine are titrated.

The effect of temperature has been described by example. The T_m shows a strong dependence on the mole fraction of guanine plus cytosine residues, F_{GC}, as well as on the cation concentration. These variables can be wed into a single empirical relationship, which, updated and refined for the conditions $0.3 < F_{OC} < 0.7$ and $0.01 < [Na^+] < 0.4$ M, is given in degrees Celsius by the Marmur-Schildkraut-Doty equation:

$$T_m = 193.67 - (3.09 - F_{GC})(34.64 - 2.83 \log_{10}[Na])$$

As noted, the stahility of DNA derives mainly from 10 neighborst-acking interactions; not just from (A·T) and (G·C) pairs. Therefore this expression represents an approximation of the more complex dependence on stacking energies; as such, it is applicable only to DNAs with suitably large and random distributions of neighbor pairs. The expression also disguises the effects of neighboring domains on denaturation, as seen in the example of Figure elsewhere in this chapter.

2.6 UTILITY OF DENATURATION STUDIES

Studies involving the deliberate denaturation of DNA are useful for many reasons, both biological and structural.

First, denaturation is necessary for producing hybrids of compleme-ntary polynucleotide chains from different source material and for forming heteroduplexes for Southern blotting or for the detection of evolutionary or mutational differences.

Second, knowledge of the precise conditions of denaturation is needed to establish a level of stringency, to ensure that chains of lesser sequence similarity are excluded from kinetically competing with complementary regions of sequence and slowing up the hybridization process.

Third, repeated cycles of denaturation and synthesis form the basis for the amplification of minute quantities of DNA by the lay-polymerase chain reaction, a simple alternative to cloning.

Fourth, the thermal denaturation of heteroduplexes is an easy and sensitive means of measuring sequence divergence, since mispairs contribute to greater configurational entropy and lower stability. The reduction of stability (i.e., the drop in melting temperature) is proportional to the frequency of' mispairs in the heteroduplex.

Fifth, denaturation is also a sensitive method for resolving distributions of sequence families of different base contents in the overall thermal denaturation profiles of total genomic DNAs.

Sixth, denaturation is required for primer bonding, an important first step in the sequencing of DNA.

Seventh, denaturing gradient gel electrophoresis is useful for the detection of point mutations and sequence variations in mixed populations of DNAs.

Eighth. denaturation of total genomic DNAs followed by careful measurements of the kinetics of renaturation is used to measure different levels of sequence complexity: the distributions of singlecopy, middle, and highly repetitive sequence families.

Finally, the thermal denaturation of DNA is an important adjunct of structural studies, providing the means for identifying various molecular sources of stability in the helix and for quantitating the thermodynamic characteristics of those sources.

3

DNA Damage and Repair

The genetic molecule DNA is continually exposed to a wide variety of chemical and physical agents that can change its structure. These changes interfere with replication and transcription of DNA and are thus referred to as DNA damage. Biological consequences include cell death and mutations, events that may cause cancers, mental retardation, and reduced growth and development.

Cells defend against the effects of DNA damage by using enzymatic repair to restore the DNA to its original structure. If repair is timely, the consequences of DNA damage are eliminated.

Obtaining an understanding of the nature of DNA damage, its biological consequences, and its elimination by repair is an important step toward resolving the many health problems associated with exposure to DNA-damaging agents.

3.1 DEFINITION OF DNA DAMAGE

Cellular modifications of DNA are induced by several agents, including electromagnetic radiation, heat, pressure, and some chemicals. Several forms of damage have been identified, but only in a few cases has their biological effectiveness been estimated. Early measures of the biological effects of radiation showed that nucleic acids were target molecules for the induction of mutations and cell death caused by ultraviolet light.

Scientists thus decided to examine the genetic molecule DNA for the chemical changes responsible. One of the first forms of DNA damage defined as biologically significant was the cyclobutyl pyrimidine

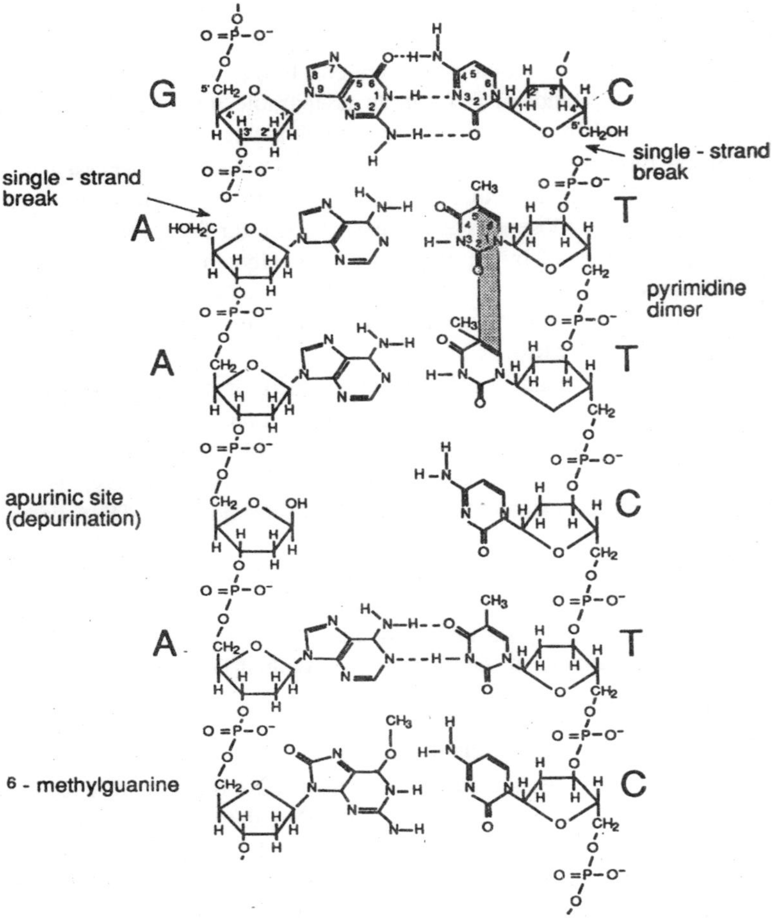

Figure 3.1: Examples of modifications that have been identified as biologically relevant DNA damage.

dimer, the result of a covalent joining of two adjacent pyrimidines in one of the DNA strands. This structure is illustrated in the representation of DNA in Figure elsewhere in this chapter.

Other examples of exogenously induced damage to DNA include breaks in the backbone structure of the DNA polymer, caused by ionizing radiation such as X-rays and by some chemicals. Heat promotes the removal of purines and pyrimidines, creating apurinic or apyrimidinic (AP) sites.

Some chemicals can alkylate the purine and pyrimidine rings of

DNA. Highly reactive radicals formed by ionizing radiation react with the purines and pyrimidines to create several oxidative products. DNA strand cross-links, which are covalent bonds that hold the two complementary strands of DNA in an inseparable state, can be created by radiation and by the action of some chemicals.

Some typical forms of the kinds of DNA damage discussed are illustrated in Figure elsewhere in this chapter. Damage to DNA induced spontaneously by endogenous agents may account for the greatest fraction of damage experienced by F most animal cells. Human body temperatures (37°C) can break the glycosidic bond between a sugar and a purine or a pyrimidine. Such action results in absent bases at sites along the DNA strands (AP sites).

Oxidants produced in cells as by-products of normal metabolism induce several DNA structural modifications. It has been estimated that these natural reactions result in 10^4 apurinic sites and an equal number of oxidant-induced damaged sites per , human cell per day. These numbers represent a small fraction (about 0.0004%) of the DNA bases per diploid cell.

However, each damaged site can cause a mutation or cell death. The recognition that normal intracellular events cause a significant level of DNA damage is a major advance in our understanding of the role of DNA damage in the health of individual cells.

3.2 DEALING WITH DNA DAMAGE

3.2.1 Biological Indications of Repair

Scientists have discovered mutant cells that are much more sensitive than normal cells to the lethal action of DNA-damaging agents. This extrasensitivity is a readily observable indicator of a defect in dealing with DNA damage. Normal cells can restore damaged DNA to its original structure; sensitive cells usually lack this capability.

3.2.2 Mechanisms of DNA Repair

Scientists have also discovered many different enzymatic mechanisms for repairing damaged DNA. These include damage-elimination by cutting it out and restoring the cut-away section and direct reversal of damage. The most common repair mechanism, referred to as excision repair, employs enzymes to remove a short section of the DNA strand containing the damage.

The opposite complementary strand is left intact. Subsequent DNA synthesis and rejoining steps fill in the resulting single-strand gap, thereby restoring the molecule to its original structure. Photoreactivation is an example of a direct mechanism for reversing UV-induced pyrimidine dimers. A specific enzyme complexed with a dimer can absorb a photon of blue light and split the cyclobutyl structure. The enzymeDNA complex then dissolves, leaving the DNA in its original unaltered structure.

3.3 DNA REPAIR IN HUMAN CELLS

3.3.1 Repair in Nomral Human Cells

Most DNA damage is removed rapidly in an error-free manner in human cells. For example, cultured human cells remove about *80%* of UV-induced pyrimidine dimers in 24 hours.

Skin cells in humans may repair the same amount of damage even faster, in 2 to 3 hours. The mechanisms for this repair are poorly understood but probably involve processes of excision and DNA resynthesis.

3.3.2 Hereditary Diseases Associated with Defects in Repair of DNA

Normal excision repair is defective in several human hereditary diseases, including xeroderma pigmentosum (XP), ataxia telangiectasia, Cockayne's syndrome, Fanconi anemia, and Bloom's syndrome. Defective repair may also cause the symptoms associated with these diseases. For example, XP patients have a high incidence of skin cancer on parts of their body that are exposed to sunlight.

Their individual cells do not repair most of the DNA damage induced by the UV portion of sunlight. This association represents some of the best evidence that DNA damage, if not repaired in a timely fashion, can induce human malignant lesions and other developmental problems.

3.4 DNA REPAIR IN AGING AND SEX

A number of theories have been proposed to account for the biological phenomena of aging and of sexual reproduction (sex). These theories can be found in volumes by Arking (aging) and by Michod and Levin (sex) listed in the Bibliography. An emerging unified theory, which accounts for much of the data relating to both aging and sex, is presented here.

Although this theory is still somewhat controversial, it is consistent with and encompasses most other theories of aging and at least one other theory of sex (see volume by Bernstein and Bernstein listed in the Bibliography).

3.4.1 Overview of the Unified Theory

Both aging and sexual reproduction (sex) appear to be consequences of a universal property of life: the vulnerability of genetic material to "noise" (DNA damage and mutation).

DNA damages occur very frequently, and organisms use enzyme-mediated repair processes to cope with them. In any cell, however, some DNA damage may remain unrepaired despite the existence of repair processes. Aging seems to be due to the accumulation of unrepaired DNA damages in somatic cells, especially in nondividing cells such as those in mammalian brain and muscle.

On the other hand, the primary function of sex appears to be repair of damages in germ cell DNA. through efficient recombinational repair when chromosomes pair during the sexual process. This mechanism allows an undamaged genome to be passed on to the next generation. In addition, in diploid organisms, sex allows chromosomes from genetically unrelated individuals (parents) to come together in a common cytoplasm (that of progeny).

Since genetically unrelated parents ordinarily would not have common mutations, the chromosomes present in their progeny should complement each other, masking expression of any deleterious recessive mutations that might be present. Thus it appears that in general, aging reflects the accumulation of DNA damage and sex reflects the removal of DNA damage plus, in diploid organisms, the masking of mutations by complementation.

3.5 THE DNA DAMAGE THEORY OF AGING

Considerable evidence indicates that DNA damage occurs frequently in somatic cells. If unrepaired, DNA damages interfere with RNA transcription and DNA replication. The result of interference, caused by accumulating DNA damage, can be progressive impairment of cell function and cell death.

Nondividing or slowly dividing cells, such as long-lived neurons and differentiated muscle cells, have been shown to accumulate DNA damage with age. In mammalian brain, mRNA synthesis and protein synthesis decline, neuron loss occurs, tissue function is reduced, and functional impairments directly related to the central processes of aging (e.g., cognitive dysfunction and decline in homeostatic regulation) occur.

Similarly, in muscle cells, mRNA and protein synthesis decline, cellular structure deteriorates, and cells die, and these processes are accompanied by a reduction in muscle strength and speed of contraction. Thus, for mammalian brain and muscle, there seems to be a good correlation between the accumulation of DNA damage and major features of aging.

In other cell populations where cell division is infrequent, such as those of liver and lymphocytes, there is also evidence that, DNA damage leads to functional declines with age. In contrast to nondividing or slowly dividing cells, at least some types of rapidly dividing cell populations seem to cope with unrepaired damages by replacing lethally damaged cells through the duplication of undamaged ones.

Such cell populations, which show little or no sign of aging, include colon and duodenum epithelial cells and hemopoietic cells of the bone marrow. Normal mammalian cellular metabolism produces oxidative free radicals from molecular oxygen, and these reactive molecules appear to cause large numbers of DNA damages per average cell per day.

It is estimated that the total number of all types of oxidative DNA damage per cell per day may be about 10,000 in man and about 100,000 in rat. Although repair processes can remove most of these damages, a significant fraction remain unrepaired in some tissues and accumulate with age. In comparisons with human, monkey, rat, and mouse, shorter life span correlates with a higher incidence of oxidatively altered DNA bases.

Thus DNA damage caused by oxidative reactions may be a major cause of aging in mammals. Support for the DNA damage theory of aging comes from some additional types of evidence. Among numerous mammalian species, life span correlates with DNA repair capacity.

Also, treatment of mammals with exogenous DNA damaging agents, such as Xrays, induces life-shortening with some similarity to normal aging. Furthermore, there are at least 10 genetic syndromes of humans in which individuals appear to have both an increased incidence of DNA damage and some features of premature aging.

3.6 REPAIR OF DNA DAMAGE

The number of DNA damages in a cell is determined by the balance between production of damages and their repair. DNA damages can be recognized by repair enzymes because of their abnormal DNA structure. This is in contrast to mutations, which cannot be recognized because they are structurally normal.

There are two basic types of DNA damage: single-strand and double-strand damages. Single-strand damages, which alter only one of the two DNA strands, can be repaired by a number of different repair processes. Excision repair, however, appears to be the predominant type used in mammalian cells. Thus excision repair may be the most important type in resisting aging.

All enzymatic pathways of excision repair involve the removal of the damaged section from one DNA strand and its accurate replacement by copying from the undamaged complementary strand. When the two strands of DNA are structurally altered at or near the same position, as occurs in a double-strand break or an interstrand cross-link, the damage is referred to as double-strand damage.

This type of damage probably cannot be repaired by excision repair, because the process depends on the ability of the DNA strand opposite the damaged one to act as an intact template for repair synthesis. Double-strand damages are formed by oxidative reactions and heat, although not as frequently as single-strand damages.

Since double-strand damages are more difficult to repair than single-strand damages, they may contribute substantially to mammalian aging despite their low frequency. To repair a doublestrand damage in one DNA molecule, it is necessary to obtain intact information from a second homologous DNA molecule.

Such repair involves replacement of the lost information by means of a transfer of a single-stranded section of the intact DNA to the damaged DNA. This process is referred to as recombinational repair. Whether recombinational repair is important in resisting mammalian aging is unknown, but the process appears to be important in sex.

3.7 THE DNA REPAIR (AND COMPLEMENTATION) THEORY OF SEX

Sexual reproduction, the most common form of reproduction in higher animals and plants, occurs among many microorganisms as well. Only about 0.1 % of higher animal species and about 8% of higher plant species entirely lack sexual reproduction. Sex, however, is a very costly process from a number of perspectives, including the time and energy females and males must spend in making contact with the opposite sex.

Furthermore, sexual reproduction, in contrast to female parthenogenesis, allows males to pass on about half their genes to each progeny while frequently avoiding the burden of nurturing these progeny. Thus a parthenogenetic female, with no male partner, can in principle often transmit her genes twice as effciently as a sexual one.

If sex is adaptive in a Darwinian sense, it must provide a large benefit to compensate for its large costs. Two adaptive benefits of sex appear to be the promotion of recombinational repair of DNA damage in the germ line, and the masking of mutations by complementation.

Recombinational repair has been shown to be efficient at overcoming DNA damages in microorganisms, especially double-strand damages. It also has been shown to be versatile (able to handle a wide variety of different types of damage) and prevalent in nature (occurring in a broad spectrum of organisms).

A key stage of the sexual cycle in eukaryotes is meiosis, the process by which germ cells, such as egg and sperm, are formed. Recombination is promoted in the phase of meiosis during which the homologous chromosomes derived from each parent line up intimately in pairs.

Gene functions necessary for meiotic recombination are required for DNA repair, and the pathway of meiotic recombination involves DNA repair steps. Meiotic recombination appears to occur by the double-strand break repair mechanism, in which a double-strand break in one chromosome is extended to a double-strand gap.

The chromosome containing the gap then receives a single-strand segment of DNA from the homologous chromosome with which it is paired. The succeeding steps involve completion of the exchange process, resulting in two intact chromosomes. This process appears to be designed to remove DNA damage in a simple, direct fashion, by creating a gap followed by doublestrand break repair at the site of the damage.

Thus meiotic recombination may reflect a general mechanism for repairing all types of DNA damage. If this view is correct, then meiosis can be regarded as an adaptation for clearing damages from the DNA passed on to progeny via germ cells. This can provide a strong selective advantage to compensate for some of the cost of sex. Outcrossing is a second key attribute of sex. This appears to be an adaptation for dealing with the problem of deleterious mutation rather than DNA damage.

For instance, humans are estimated to carry (in a heterozygous state) an average of about two deleterious recessive mutations per genome. Such mutations can impose a large physiological penalty on inbreeding. That is, incestuous coatings in humans often result in genetically defective progeny.

In an organism having a diploid stage in its life cycle, wild-type (functional) genes in one copy can complement (mask) mutated genes in the other copy present in the same cell. However complementation is most beneficial if the mating partners are unrelated, since their genomes are unlikely to carry the same deleterious recessive mutations. Thus complementation that occurs upon outcrossing provides a further strong selective advantage in diploids that could also bompensate for some of the high cost of sex.

In conclusion, sex seems to be maintained in diploid organisms both by the advantage of recombinational repair during meiosis and the advantage of outcrossing, which facilitates complementation of recessive deleterious mutations.

It appears that the need to cope with genetic noise (DNA damage and mutation) in the transmission of genetic information has generated a continuous selective pressure for sex. In humans, the availability of efficient repair during meiosis may underlie the potential immortality of the germ line. In contrast, the less efficient DNA repair available in the somatic cells may account for the aging and mortality of the individual.

4

DNA Sequencing and Finger Printing

DNA sequencing by horizontal ultrathin gel electrophoresis (HUGE) is a method of separating DNA in thin (usually < 0.1 mm) polyacrylamide gels so that the relative size of the DNA fragments (and thus, the sequence) can be determined. When used with conventional radioisotope-based sequencing protocols, HUGE can resolve up to 350-400 bases of sequence information in less than 30 minutes of electrophoresis, instead of the several hours usually required by traditional vertical electrophoresis.

HUGE is also compatible with automated DNA sequencing methods and reduces the time of analysis by about a factor of 10. DNA separation by HUGE is a relatively new technique that is gaining acceptance as a practical way to significantly decrease the time required for the electrophoresis step of DNA sequencing.

4.1 PRINCIPLES

At present, almost all DNA sequencing is based on the separation of DNA fragments in polyacrylamide gels. One approach to reducing the time required for this step is to increase the speed of the separations. Several researchers have shown that capillary gel electrophoresis permits the time required for the electrophoretic separation of DNA fragments to be greatly reduced.

Capillary gels are only 50 μm in diameter and are very efficient at dissipating the heat generated in electrophoresis. Thus, electric field strengths as high as 400 V/ cm can be used for DNA separation in capillary systems without excessive Joule heating. The use of high electric fields can result in separation speeds increased as much as 26-

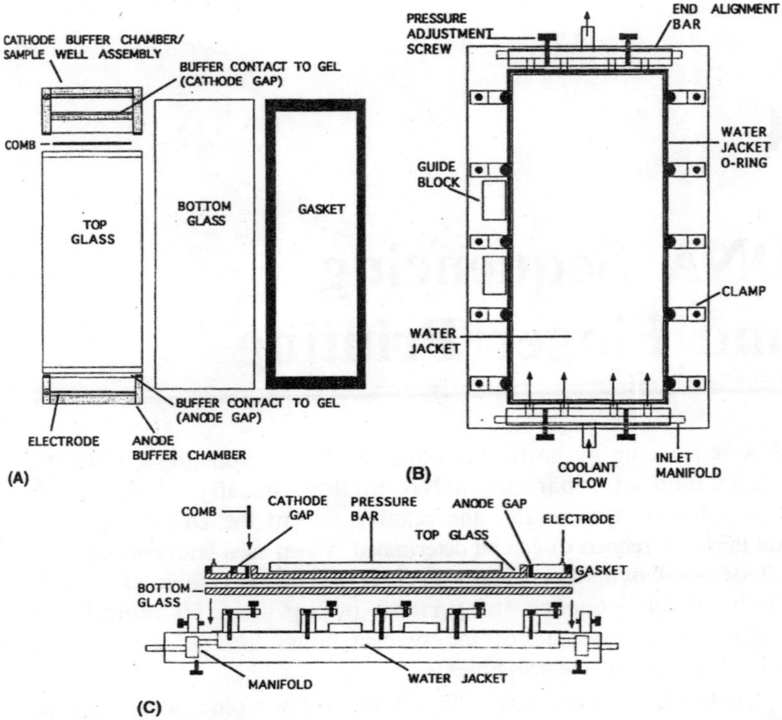

Figure 4.1: The HUGE setup. (A) Top view of glass and plastic components. (B) Top view of horizontal apparatus base. (C) Side view of horizontal apparatus base and glass components.

fold over conventional electrophoresis. However, for each DNA sample that is analyzed in parallel by capillary gel electrophoresis, one gel-filled capillary must be prepared.

Our approach to achieving rapid separations in polyacrylamide gels has been to develop a method of introducing samples onto 50 Rm slab gels. We describe here the instrumentation and methods used for high speed DNA sequencing by horizontal ultrathin gel electrophoresis (HUGE).

4.2 APPARATUS DESIGN

Schematic diagrams of the apparatus are shown in Figure 1. The two major subdivisions of the horizontal apparatus are the glass and plastic "components" and the apparatus "base". The "components," when assembled, define the gel dimensions, sample wells, and electrophoresis buffer chambers. The "base" provides a water jacket, guide blocks, and clamps to hold the "components" in place. The temperature of the gel is regulated during electrophoresis by circulating coolant through the water jacket under the glass plates.

The manifolds help to disperse the fluid across the width of the glass and aid in the uniform distribution of temperature across the gel.

4.3 TECHNIEQUES

4.3.1 Gel Preparation and Assembly

Prior to gel preparation, all glass components are cleaned and the bottom glass is treated so that the gel will bond to the glass when the "components" are disassembled after electrophoresis. The gel is cast directly on the apparatus by assembling the "components" as shown in Figure elsewhere in this chapter.

When these units have been assembled and clamped into place, the polyacrylamide solution is poured into one of the buffer chambers and the gel flows down the glass within the inside border of the gel gasket. The three vertical "gaps" (comb, anode, and cathode) arc filled with gel solution, and the comb is placed in the slot between the cathode chamber and top glass plate. The gel is allowed to polymerize for about 60 minutes before use.

4.3.2 Sample Loading and Electrophoresis

About 15 minutes before samples are to be loaded, the water in the circulating bath is preheated to the desired temperature (usually about 35-40°C for sequencing gels). After preheating, the water is allowed to circulate through the water jacket via the inlet and outlet manifolds, to prewarm the glass plates.

The comb is then removed, the sample wells are rinsed with distilled water, and sequencing reaction products are dispensed into the wells with a 1.0 p.L syringe. The electrodes arc connected from the power supply to the apparatus, and electrophoresis buffer is added to the anode and cathode chambers.

Gels are usually electrophoresed at 40-50 W constant power, which results in a field strength of 150-250 V/cm. Electrophoresis time varies, depending on the amount and location of the sequence of interest. If the first 300 bases of sequence are desired, electrophoresis is stopped after about 15-20 minutes, when the bromophenol blue tracking dye reaches the anode buffer chamber (this dye is usually included in the "stop" solution supplied with many commercial radioactive sequencing kits).

When electrophoresis is complete, the buffer is removed from the chambers and properly discarded (the buffer in the anode chamber is radioactive). The water circulator is turned off and the water jacket is drained of fluid. The "components" are removed from the apparatus base and disassembled such that the gel adheres to the bottom glass plate.

The gel can then be "fixed" with an alcohol/acid solution and dried or wrapped in plastic wrap and exposed to X-ray film to visualize the separated fragments. Figure elsewhere in this chapter is an example of a typical sequencing "ladder" produced by the Sanger dideoxy radioactive sequencing protocol.

4.4 AUTOMATED DNA SEQUENCING

The speed of DNA sequencing can be increased further by using fluorescent-labeled DNA instead of radioactive labels. The rationale and methodology of this approach have been reviewed extensively. We have developed a detection system that is compatible with HUGE in which 450 bases of sequencing information can be obtained from each of 18 samples in about one hour.

This system consists of a laser and optical components to provide a line of fluorescence excitation across the gel, collection optics, and a charge-coupled device (CCD) detector to measure the fluorescence emitted from the DNA samples. The detection system is connected to a computer that controls the CCD camera functions and stores and analyzes the data.

The output of data appears on the computer screen as a series of peaks, each peak corresponding to a fragment of DNA. This automated system is currently under development to optimize its mance and promises to substantially increase the speed and the cost of DNA sequence analysis.

4.5 DNA FINGER PRINTING ANALYSIS

DNA fingerprint analysis (fingerprinting or typing) is the process of preparing and interpreting bar-code-like profiles of DNA segments for individual identification. Nanogram-to-microgram quantities of genomic DNA are isolated and polymorphic segments are directly analyzed or amplified by means of the polymerase chain reaction (PCR) and analyzed.

With the exception of identical twins and clones, profiles of the segments (usually separated and characterized according to length or by sequence differences) from humans, other animals, microorganisms, fungi, and plants are unique to each individual.

The analysis of DNA is revolutionizing the field of identification. The need for only minute quantities of tissue, the stability of DNA, and the high degree of assay accuracy and precision have contributed to the universal application of DNA typing.

The use of this procedure is rapidly altering the manner in which genetic analyses are carried out in human parentage, rape, and homicide cases; in animal poaching, parentage, breeding, and population studies: in medical analysis; and in plant patent disputes, clone identity investigations, parentage testing, and gene bank management programs.

4.6 PRINCIPLES

4.6.1 Specimen Processing

DNA is the molecule of heredity. It is an integral component of all living matter with the exception of RNA viruses. Any living or nonliving organic matter containing relatively intact DNA fragments can be used for DNA typing analysis.

Common sources include human or other animal blood, semen and solid tissues, and plant foliage and seeds. A document detailing chain of custody (continuity) is required for a legal case specimen. The drawing and labeling (name ID, date, location) of blood samples in paternity and immigration cases must be witnessed, as must details of specimen collection and custody in rape and homicide cases.

Specimens are stored under clean, dry, cool conditions to reduce

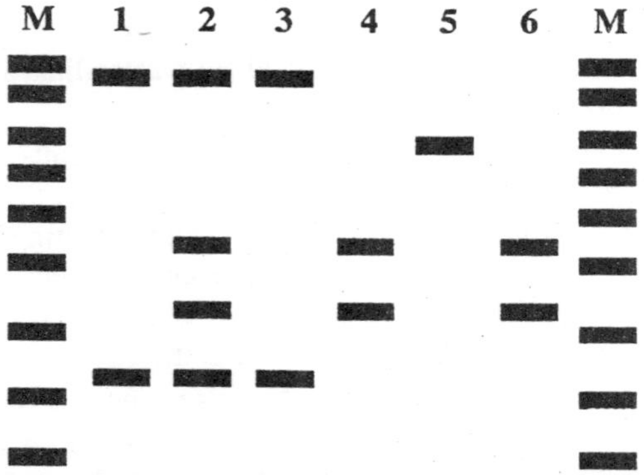

M Molecular Weight Marker
1 Victim control standard
2 Vaginal swab mixed sample
3 Differential extraction of swab: female fraction
4 Differential extraction of swab: male fraction
5 Suspect #1 control sample (exclusion)
6 Suspect #2 control sample (inclusion)

Figure 4.2: DNA typing in a sexual assault case.

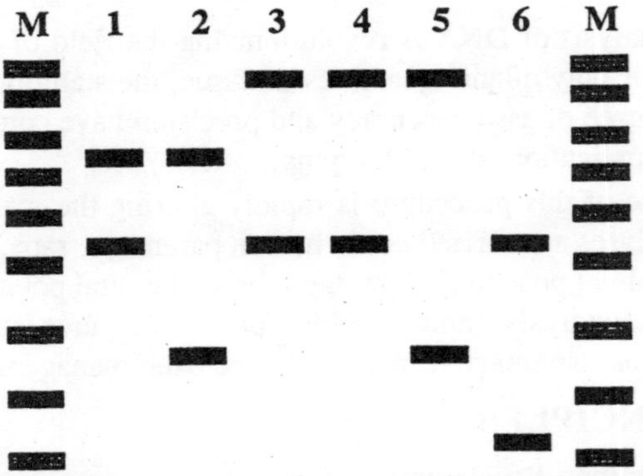

M Molecular Weight Marker
1 Mother
2 Child (paternity inclusion)
3 Child identical twin to 4 (paternity inclusion)
4 Child identical twin to 3 (paternity inclusion)
5 Father
6 Child (paternity exclusion)

Figure 4.3: Paternity analysis by DNA.

contamination by microorganisms and DNA degradation caused by cell lysis and DNAase activity. Successful analyses have been carried out on DNA extracted from dried blood stains, dried museum specimens, and frozen tissue preserved for many decades. Indeed, these studies have formed the basis of recent new theories in molecular evolution and anthropology.

A good yield of high molecular weight DNA devoid of organic and inorganic contaminants can be extracted from tissues. The enzyme proteinase K is generally used to assist with cell lysis and to digest proteins. The detergent sodium dodecyl sulfate facilitates the separation of residual protein from DNA.

Proteinaceous materials are removed by phenol extraction (or a nonorganic NaCl or LiCl precipitation), and chloroform is added to remove traces of phenol. Most of the contaminating RNA can be eliminated, provided the phenol is appropriately equilibrated. High molecular weight DNA is recovered in a relatively pure form by ethanol precipitation in the presence of salt.

The extraction and purification steps can, under certain circumstances, be bypassed after cellular lysis. Fragments of DNA can be directly amplified by PCR and the amplified fraction used to construct a DNA fingerprint profile. If this approach is followed, possible inhibitors of the PCR reaction can be removed by means of a simple nonorganic technique using an ion exchange resin such as Chelex 100 or by washing through a Centricon 100 filter.

The analyst may be presented with mixed specimens. It is possible to separate mixtures such as sperm and female cells often present in vaginal swabs from rape victims. The female cells are lysed in extraction buffer devoid of dithiothreitol (DTT) and the DNA is isolated from the supernatant. The sperm pellet is then lysed by adding extraction buffer containing DTT and increased detergent.

DNA quality can be evaluated by ethidium bromide staining or by southern blotting and hybridizing with a species specific probe. The degree of degradation is estimated by observing, under UV light, ethidium bromide stained fragment lengths subjected to agarose gel electrophoresis.

High molecular weight DNA appears as a single large band, whereas partially degraded DNA forms a long smear of large to small fragments (> 20,000 base pairs to a few hundred base pairs for human DNA). Ethidium bromide staining cannot differentiate DNA from different organisms-speciesspecific probes are used for this purpose.

Quantitation is the final step in preparing DNA for analysis. At least three different procedures are available. In the first technique, the absorbance of a DNA solution is measured in a spectrophotometer at 260 nm wavelength. Second, the DNA and a series of standards of known concentration are subjected to agarose gel electrophoresis and stained with ethidium bromide. Last, an aliquot of the DNA is fixed on a nylon membrane and is hybridized with, for example, a labeled, highly repetitive, primate-specific a-satellite probe pl7H8 that detects locus D17Z1.

4.6.2 Specimen Analysis

The analysis objective is the isolation of DNA segments that will facilitate identity discrimination among individuals. Profile patterns from tissues from the same individual should be identical. Many polymorphic genetic markers (segments) consist of tandem repetitive sequences.

Depending on the locus, each repeat can consist of a minimum of 2 to perhaps 80 base pairs and may contain possible single base pair polymorphic sites. The number of tandem repeats varies from one to hundreds. A locus may, therefore, be polymorphic for the number of tandem repeats [referred to as variable number (VNTRs) or short (STRs)] and for the base pair composition at specific sites within each repeat [minisat-ellite variant repeat (MVR)].

This variation, when analyzed at one or more loci by RFLP, AMP-FLP, or MVR-PCR techniques is the basis of DNA profile construction. Polymorphic sites that may or may not be located in tandem repeat units provide another form of variation useful in identity analysis. The

RAPD, AP-PCR, or DAF system is based on the annealing of arbitrary sequence PCR primers to template sequences sufficiently close together on complementary DNA strands to facilitate amplification.

Systems differ in their specifics: the length of the primers used, the amplification conditions, and the resolution an isualization of the products. Amplification will not occur if the template sequence differs significantly from the primer sequence or perhaps if competition occurs between the primer and hairpin loop structures in the template.

The final DNA pattern or profile can be visualized with colour dyes, fluorescent dyes, radioisotopes, or other stains such as silver. There are three analytical approaches used for DNA fingerprinting: VNTR analysis, dot blot analysis of sequence polymorphisms, and direct sequence analysis.

For VNTR analysis, the restriction-digested or -amplified DNA fragments are separated, generally based on size, by electrophoresis in a sieving medium. The relative positions of the bands after electrophoretic separation are determined with reference to size standards. In dot blot analysis, the target DNA sequence is amplified by PCR and the amplified products are fixed to membranes.

The DNA is then incubated with allelespecific probes. If hybridization occurs, the dot will be marked; if the DNA does not contain the particular allele, the spot will remain blank. Alternatively, dot blots can be configured where the allelespecific probes are fixed to the membrane and the amplified DNA is allowed to hybridize to the fixed probes.

This reverse dot blot approach is used for HLA-DQa typing. Finally, the sequence of DNA bases can be read by direct sequencing. A number of analytical approaches can, therefore, be used to isolate and identify profile patterns. Isolated and purified DNA or nonextracted DNA may be used. DNA from each of a number of loci may be analyzed separately and the results combined to form a profile, or DNA consisting of similar base pair sequences from multiple loci may be analyzed coincidentally. The DNA may be analyzed directly, or it may be amplified and then analyzed.

4.6.3 Data Processing

Genetic profiles are presented as (1) the presence or absence of dots, (2) the position or size of DNA fragments separated by an electrophoretic approach, or (3) direct sequence information (e.g., AATCGTACCTGATCC). The results produced by any of these methods can be evaluated visually. However, as data become more abundant and

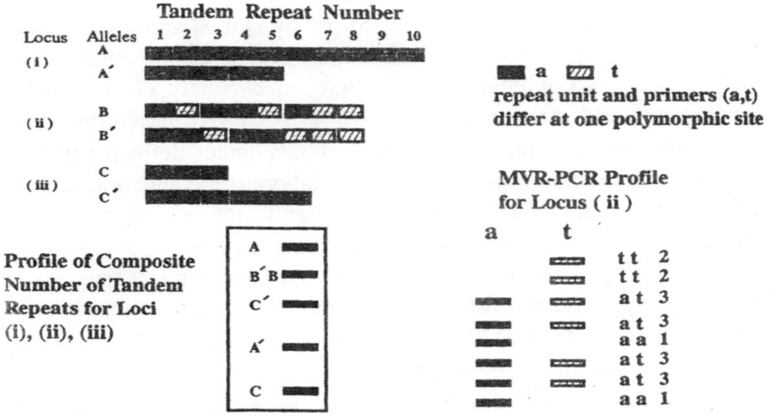

Figure 4.4: Variable number of tandem repeats and MVR-PCR profiles.

complex, computer-assisted analyses are needed to aid in pattern interpretation, data storage and manipulation.

For dot blot profiles, most interpretations are performed by "eye," and the data are generally transcribed manually into a computer storage bank. However, since most dots are coloured or visible on X-ray film, densitometric scanning systems can be used to automatically score and store the data for future analysis.

For most VNTR analyses, which are based on size or position, the first generation of data acquisition used to determine the relative size of unknown DNA fragments was processed by measurement with a ruler and comparison with known size standards (on autoradiograms). Later, semiautomated computer-assisted image capture systems (which are essentially electronic rulers) replaced the manual method of data analysis.

With the use of fluorescent or UV detection systems, labeled DNA fragments can be detected in realtime analysis and automatically transcribed to the computer. Both VNTR typing and sequencing can be automated using this approach.

The main advantages of nonmanual data acquisition are the reduction of human transcriptional errors and, hopefully, a reduction in labor. Once the data have been stored in the computer, various simple and sophisticated statistical analyses can be performed for determinations, such as the likelihood of occurrence of DNA profiles for identity typing purposes and for linkage analysis.

4.6.4 Quality Assurance

Laboratory quality assurance is the documented verification that proper procedures have been carried out by skilled and highly trained personnel to ensure that valid and reliable results can be obtained. The validity of DNA typing results centers on correctly identifying true nonmatches and matches from a prospective DNA fingerprint profile.

Test reliability measures reproducibility under defined conditions of use and should transcend different laboratories and practitioners. Quality assurance must encompass all significant aspects of the DNA typing process, which include personnel education and training, documentation of records, data analysis, quality control of reagants and equipment, technical controls, proficiency testing, reporting of results, and auditing of the laboratory procedures.

Quality assurance and appropriate standards for DNA typing evolved initially from the experience of clinical laboratories; more recently, however, they have been carefully defined through consensus from forensic laboratories represented in the North American nonregulatory federal initiative, the Technical Working Group on DNA Analysis Methods (TWGDAM).

It is important to note that quality assurance guidelines in forensic science must retain enough flexibility to accommodate the unique nature of forensic samples as well as future advances in recombinant DNA technology and molecular biology. Good quality control management is paramount for obtaining quality results. It is also a key factor in establishing uniform reliability between laboratories.

4.7 APPLICATIONS

DNA fingerprinting is applicable to any identification problem for which a minimal quantity of relatively intact DNA is available. The requirement for intact DNA depends on the method of analysis. Typing by means of restriction fragment length polymorphisms (RFLP) generally requires high molecular weight DNA, whereas most genetic marker typing based on PCR can make use of much smaller, even somewhat degraded, fragments.

4.7.1 Human

Suspect or victim tissue found at the scene of a rape or a homicide provides a source of forensic evidence for determining whether a suspect is the source of the crime scene specimens. Accident or homicide victims unidentifiable from their physical features can be identified provided typable DNA is isolated from the victim's remains and is matched with DNA, perhaps from hair roots obtained from his or her hair brush.

If the victim's putative parents or other relatives are available, parentage (or other biological relationship) testing also may be carried out. Parentage can be determined for child custody and immigration cases, and for counseling about genetic diseases. The bands in the offspring's DNA profile must have been inherited from the biological parents. Barring mutations, the presence of bands not found in either parental profile is indicative of nonparentage.

Applications in medicine include genetic counseling, tracing the percentage of donor versus recipient cells in bone marrow transplants, determination of possible "contamination" of fetal chorionic villi sample (CVS) tissue with maternal tissue, tissue culture cell line identification, and the confirmation of twin zygosity.

Identification monitors are feasible for security purposes in the armed forces and for identification in mass disasters or war. For these applications, DNA profiles are recorded for future match comparisons.

4.7.2 Animal

Parentage, poaching, identification in cases of theft or loss, population studies, the detection of trait markers, and breeding programs are areas of application in animal husbandry. Proof of parentage may be important for pedigree registration and for establishing sale value. A DNA profile from animal remains can be compared with a profile from a frozen steak or from a trophy specimen in a poaching case.

The evolution of populations can be traced, for example, by analyzing DNA from dried museum specimens or from insects housed in amber for millions of years. Base pair sequences and profile bands can be compared with those from modern relatives.

Profiling is used in breeding programs for populations of endangered species with small gene pools. Animals exhibiting the greatest number of DNA profile differences and, therefore, probably the greatest genetic diversity, are chosen for breeding. Also, biological relationships can be confirmed in artificial insemination and embryo transfer programs for domesticated animals.

4.7.3 Microorganisms

DNA fingerprinting provides a powerful tool for microorganism identification. Different bacterial strains, for example, have unique profiles. Profiling also provides a basis for issuing patents for microorganisms such as yeast used in the brewing industry.

4.7.4 Plants

The field of plant identification for patent, parentage, theft, and trait marker purposes has recently received considerable impetus because of DNA analysis, especially using the RAPD-type systems.

The infringement of breeders' rights relating to the origin of cloned material such as cuttings, parentage of plants with desirable traits, the theft of expensive trees, and gene bank management for identification and the measurement of variation are examples of applications.

4.8 PERSPECTIVES

4.8.1 DNA Fingerprinting versus other Identification Techniques

Many of the conventional biochemical identification techniques such as isozyme electrophoresis are being replaced by DNA typing, especially where more information is required. Although conventional methods such as blood typing are rapid, simple, and definitive for exclusions, they are much less powerful than DNA analysis for the probability of inclusion.

If a mother and a putative father both have blood type 0, their offspring cannot be type A, B, or AB. The putative father is readily excluded as the biological father in this example. If the offspring is type O, possible exclusion of the putative father is considerably more difficult using conventional methods.

Additional key factors in choosing DNA for analysis include the availability of the hereditary material, as well as its highly polymorphic nature, continuity within the same organism, stability relative to that of enzymes, and ability to amplify segments in vitro when a minimal quantity of tissue is available or the DNA is moderately degraded.

4.8.2 Future Directions

Automation is a primary objective. Current methodology is labor intensive because of the requirement of a number of isolated steps, including combinations of extraction, amplification, endonuclease digestion, electrophoresis, labeling, hybridization, blot scanning, profile comparison matching, data reduction, and probability calculation.

Automation is especially desirable for direct sequencing of highly polymorphic regions of the genome because it will result in the elimination of most of the tedious current techniques. Although fingerprinting approaches using RAPD (DAF, APPCR) systems circumvent the need for probes, a limitation arises in the calculation of probabilities.

It is extremely difficult to determine population frequencies of the amplified fragments. Without these data, the calculation of the probability of profile matches is not readily feasible. Probes (together with their population allele frequencies) specific to each species are, therefore, urgently needed. DNA fingerprinting began in 1985 with multilocus analysis of sperm and blood in a double rape-homicide case in Britain.

In less than a decade, applications have reached into every taxonomic kingdom-Monera, Protista, Fungi, Animalia, and Planta. It appears that the potential uses of molecular biology techniques for identity testing have only begun to be realized.

4.9 DNA MARKERS, CLONED

It is possible to clone (isolate) any DNA fragment in the genome of an organism and, after reverse transcription, any transcribed gene in the form of a cDNA. The cloning procedure involves the insertion of the DNA fragment into a vector, capable of replication in a microorganism; it is then possible to produce a large quantity of the DNA fragment for physical or biological analysis.

Determination of the location in the genome from which the particular DNA fragment was derived confers on that fragment the property of a DNA marker. Such DNA markers are a prerequisite for physical and genetic mapping of the genome of the organism.

DNA markers are also of importance for the diagnosis of genetic diseases. DNA markers can be divided into several different classes. Vectors and clone libraries of different types can be used.

4.9.1 Principles

In molecular biology, cloning is the insertion of DNA with interesting information into a specific vector that allows replication and transfer of the cloned DNA from one host to another. The clone containing the inserted DNA is called a "recombinant vector" to distinguish it from its parental vector, which does not contain any inserted foreign DNA.

The main idea of cloning is to obtain the interesting piece of DNA in a large quantity convenient for analysis and further experiments. Now there are many different types of vectors and many strategies used for cloning.

This entry concentrates on the widely used lambda-based vectors and the construction of different types of genomic libraries. A genomic library is a collection of recombinants containing DNA fragments that represent the genome of a particular organism.

Genomic libraries can be either general, containing DNA fragments covering the whole genome, or special, containing only specific genome fragments, which differ in certain parameters. Cloned DNA fragments can be located to a specific site of a chromosome, whereupon they can serve as markers for physical and genetic mapping. Anonymous markers represent randomly cloned DNA fragments whose functions or specific features are not known.

Other DNA markers can contain a known gene, recognition sites for rare cutting restriction enzymes, and so on. Such markers can be polymorphic, that is, having different structures in different individuals (they are usually distinguished on the basis of differing mobility in gel electrophoresis). Polymorphic markers come in three common types. Restriction fragment length polymorphism (RFLP) markers recognize genomic fragments that contain polymorphic recognition sites for a particular restriction endonuclease (epg., *Tay* I, Msp I).

The same chromosomal regions in different individuals contain or lack this recognition site. Another form of DNA polymorphism results from variation in the number of tandemly repeated (VNTR) DNA sequences in a particular locus. Usually the VNTR sequences are divided in two classes: mini- and microsatellites.

Minisatellites are DNA fragments 0.2 to 2 kilobases long that contain many copies of 15 to 60 by repeats. Microsatellites are relatively short DNA fragments (usually <I 00 bp) that contain runs of tandemly repeated DNA with repeat unit 1 to 5 bp.

4.10 TECHNIQUES

4.10.1 General Characteristics of Lambda-Based Vectors Used Construction of Genomic Libraries

Extensive modifications of lambda phage vectors were developed which combine the features of different vector systems. Cosmids are essentially plasmids that contain the cos region of phage lambda responsible for packaging of DNA into the phage particle.

Since the plasmid body is usually small (3-6 kb), big DNA molecules (46-49 kh) can be cloned in these vectors. Phagemids are lambda phages that have an inserted plasmid. Diphagemids are the vectors that combine the advantages of phages (X and M13) and plasmids.

4.10.2 Construction of General Genomic Libraries

Representativity is one of the most important features of a genomic library. In a representative genomic library, every genomic DNA fragments will be present in at least one of the recombinant phages of the library. Two common ways to construct genomic libraries are shown in Figure elsewhere in this chapter.

The first method includes generation of sheared genomic DNA

fragments of a particular size. The vector DNA is digested with two restriction enzymes. The stuffer piece and both arms will now have different sticky ends, preventing re-creation of the original vector molecule during subsequent ligation with genomic DNA fragments.

The "partial filling-in" method avoids fractionation steps. Phage arms are prepared as described before, and the sticky ends produced are partially filled in with the Klenow fragment of DNA polymerase I. Genomic DNA partially digested with Sau3AI is also partially filled in, but in the presence dATP and dGTP. Under the preceding condition, self'-ligation of vector arms or genomic DNA is impossible.

4.11 APPLICATIONS AND PERSPECTIVES

4.11.1 Cloning DNA Markers Specific for a Particular Chromosome

One approach to cloning DNA markers that are specific for a given chromosome is based on the use of fluorescence-activated cell sorters (FACS). DNA isolated from sorted chromosomes can be used for construction of chromosome-specific genomic libraries. Another approach is based on the use of hybrid cell lines.

To obtain such somatic cell hybrids, human cells are fused with rodent cells. When the resulting hybrid cells are grown in culture, there is a progressive loss of human chromosomes until only one or a few of them are left. Human-specific clones can be isolated from the library by hybridization to total human DNA.

4.11.2 ALU-PCR as a Tool to Clone Marker from Specific Regions of the Chromosomes

Alu repeats are the most abundant repetitive sequences in the human genome. Related repeats exist also in other mammals. It is possible to find conserved sequences that are species specific. These conserved sequences can be used as primers in the polymerase chain reaction (PCR), to specifically amplify human sequences in the presence of nonhuman DNA.

These features are the basis for using *Alu* repeats for isolation (and cloning) of human chromosome specific sequences from hybrid cell lines containing human and nonhuman DNA sequences.

4.11.3 Use of Microdissection to Construct Region-Specific Libraries

Another approach to obtaining region-specific libraries is chromosome microdissection. This approach makes it possible to physically remove the chromosomal region of interest and to clone minute quantities of microdissected DNA by a microcloning procedure. DNA obtained from only a few chromosomes is enough for PCR amplification and construction of a regionspecific library.

4.11.4 CPG Islands as Powerful Markers for Genome Mapping

Stably unmethylated sequences (about 1% of the genome) have been observed in human chromosomal DNA. They are usually called CpG (rich) islands because they contain more than 50% of C+G. It is now clear that the majority if not all CpG islands are associated with genes.

It has been shown that recognition sites for many of the rare cutting enzymes are closely associated with CpG islands (e.g., 82% of all Not I sites are located in the CpG islands).

4.11.5 Linking and Jumping Libraries

The Not I jumping library and the "general" jumping (hopping) library are the two best-known kinds; only the former is discussed here. The basic principle of both methods is to clone only the ends of large DNA fragment rather than continuous DNA segments, as in yeast artificial chromosome (YAC) clones.

Internal DNA is deleted by controlled biochemical techniques. Jumping clones contain DNA sequences adjacent to neighboring Not I sites, and linking clones contain DNA sequences surrounding the same restriction site. There are two main approaches for the construction of Not I jumping and linking libraries.

In the first method, jumping and linking libraries are constructed using selective genetic marker. An integrated approach for construction of jumping and linking libraries is outlined in Figure elsewhere in this chapter.

4.11.6 ALU-PCR and Subtractive Procedures to Clone CPG Islands from Defined Regions of the Chromosomes

The disadvantage of the *Alu-PCR* approach described in other section of this book is that the DNA fragments cloned are small and not linked with genes or markers of other kinds. According to another scheme, linking libraries are constructed from different hybrid cell lines containing either whole or deleted human chromosomes.

Then total DNA isolated from these libraries can be used for Alu-PCR. Genomic subtractive methods represent potentially powerful tools for identification of deleted sequences and cloning region-specific markers. But the great complexity of the human genome has generated serious problems. These problems can be overcome by reducing the complexity of the human genomic sequences. Two approaches have been suggested to achieve this aim.

In one (representational difference analysis), only a subset of genomic sequences (epg., BarnHI fragments / kb) is used for subtractive procedures, and this approach will result in the cloning of random sequences. In the other, Not I or Xho I (Sal I) linking libraries are used instead of whole genomic DNA. It is important to note that linking clones are associated with genes.

4.11.7 Use of Linking and Jumping Clones to Construct a Physical Map of Chromosomes

When linking clones are used to probe a genomic blot (Not I digested genomic DNA) produced by pulsed field gradient electrophoresis, each clone should reveal two DNA fragments, which are adjacent in the genome.

Thus, in principle, one should he able to order the rare cutting sites with just a single library and one digest. although it seldom is possible to distinguish between two fragments of the same size. The use of jumping and linking libraries in a complementary fashion simplifies this approach.

5

Structure of RNA

Many functions of biological macromolecules require adoption of a single, three-dimensional structure that is recognized by other biomolecules. While hundreds of three-dimensional structures of proteins are known, only a few RNA structures have been determined.

Moreover, the most functionally important parts of RNA-the irregular, nonhelical loops-are also the least well understood in structural terms. Determination of nonhelical RNA structures will improve attempts to predict RNA secondary and tertiary structure, provide insights into the specificity of proteinRNA recognition, and shed light on the mechanism of RNA-mediated catalysis.

5.1 NUCLEIC ACID STRUCTURE

The elements of regular secondary structure in nucleic acids and proteins-the double helix, the a helix, and the 1 sheet-were all discovered in the early 1950s. Today, however, knowledge of protein structure has progressed far beyond that of nucleic acid structure. While hundreds of three-dimensional protein structures are known, the number of nucleic acid structures that have been determined to high resolution (most of which have been fully Watson-Crick-paired double helices) is much lower.

Thus especially little is known about nonhelical nucleic acid structure, which is unfortunate because it is often these loop structures that are most important to function. Why have nonhelical DNA and RNA structures been so difficult to determine'? For nucleic acids, nonhelical means irregular, and irregular conformations depend much more on sequence than regular, helical ones.

The determination of nonhelical nucleic acid structure therefore requires nucleic acids of defined length and sequence, and experimental techniques that can provide conformational details at atomic resolution.

Because only a few of the smaller naturally occurring nucleic acids can be isolated from natural sources in sufficient, pure quantities for physical studies, most of the advances in the study of nonhelical nucleic acid structure had to await improved methods for oligonucleotide synthesis.

In addition, only X-ray crystallography and very recent NMR techniques allow determination of atomic resolution structures, and only a very few nonhelical nucleic acids have been successfully crystallized. However, improved methods are now available for synthesis, crystallization, and isotopic labeling for NMR studies, which should accelerate the determination of nonhelical nucleic acid structures.

As more irregular structures are determined, new rules for their Iblding will be discovered, which will improve the prediction of nucleic acid secondary and tertiary structure given only the RNA or DNA sequence. This contribution focuses on high resolution nonhelical RNA structures.

While nonhelical conformations are required for some DNA functions, they are relatively rare and often transient structural features. Nonhelical RNA structures, on the other hand, are common, persistent, and crucial to function: typically it is the irregular RNA loops rather than the helices that bind protein, nucleic acid, or ion ligands.

The structures of three transfer RNA anticodon hairpin loops are described in detail to illustrate emerging principles of nonhelical RNA structure. Special emphasis is placed on the nonWatson-Crick interactions that stabilize irregular conformations.

Before delving into the details of three-dimensional structures, we briefly review the nomenclature and physical forces that determine RNA conformation.

5.1.1 Nomenclature of RNA Structure

Unlike DNA, which usually consists of two fully Watson-Crickpaired strands, most RNA is found as a single strand, normally a linear chain of the four nucleotide monomers adenine (A), guanine (G), cytosine (C), and uracil (U).

Each nucleotide monomer consists of a side chain (a heterocyclic base) bonded to the sugar ring of a sugar-phosphate backbone. Note the presence of the ribose sugar 2′-OH group, which is an H atom in DNA. The nucleotide units are linked head-to-tail, with the 03′ atom of residue i bonded to the phosphorus atom of residue i+1.

This configuration of the RNA chain imparts a directionality to it, so that, for example, 5′ AGG 3′ is not the same molecule as 3′ AGG 5′. The threedimensional structure of an RNA molecule is fully determined by specifying for each nucleotide seven torsion angles describing orientations about rotatable single bonds. Six of these torsion angles (α, β, γ, δ, ε, and ξ) describe the conformation of the backbone atoms (Figure 1, in white), while one torsion angle (X) describes the orientation of the base with respect to the ribose sugar.

Angle a describes the orientation about the P—O5′ bond, β that about the O5′—C5′ bond, and so on, with ξ (O3′—P) at the end of the nucleotide unit. While this description of RNA structure is very useful for classification of different loop structures, it says nothing about the specific interactions that stabilize the three-dimensional structure.

5.1.2 Physical Forces Determining RNA Conformation

The two most important noncovalent forces affecting nucleic acid conformation are hydrogen bonding and stacking of the aromatic bases. The environment, as determined by interactions with proteins, water, or counterions, is also of great importance, but a detailed description of intermolecular interactions with RNA is beyond the scope of this contribution.

Crystallographic studies of oligonucleotides and transfer RNAs reveal that owing to shortrange dipole-induced dipole forces, the bases are almost always well stacked. The energy associated with such stacking interactions depends on the identity of the bases and the detailed geometry of the interaction; generally the purine (G and A) bases have a greater tendency to stack than the pyrimidine (C and U) bases.

Hydrogen bonds are divided into three categories: base-base, base-backbone, and backbone-backbone. The geometry of Watson-Crick-paired helices allows direct hydrogen bonding between base functional groups (base-base) only. The irregular backbone conformation of loops can allow, in general, close approach of all functional groups, so that all three categories of hydrogen bonds are found.

Another kind of backbone-backbone interaction is the close approach of phosphate groups, which may create an ion-binding site. In other section of this book we describe the structures of several RNA hairpins in detail, to illustrate specific examples of such interactions and some of the recurrent structural motifs found in nonhelical structures.

5.2 A COMPARISON OF THREE TRANSFER RNA ANTICODON HAIRPIN LOOPS

The differences in conformation between the anticodon loops from three different transfer RNAs illustrate the flexibility of large hairpin loops. The loop at upper left is from the $tRNA^{Asp}$-cognate synthetase cocrystal structure, that at upper right is from the $tRNA^{Gin}$-cognate synthetase cocrystal structure, and that at bottom is from the free yeast $tRNA^{Phe}$ crystal structure.

The conformations of the free $tRNA^{Asp}$ and $tRNA^{Gln}$ loops are similar to that of the $tRNA^{Phe}$ loop. Therefore, the differences in conformation between the protein-hound loops and the uncomplexed $tRNA^{Phe}$ structure may be interpreted as conformational changes that occur upon binding of the synthetase.

5.2.1 $tRNA^{Phe}$

The $tRNA^{Phe}$ anticodon loop has long been considered to be the canonical seven-residue hairpin loop. The structure features helical stacking of the five bases (residues 34-38) at the 3′ side of the loop, and a tight turn motif involving residues 33-36, where the direction of the backbone is reversed [note the difference in orientation of the U33 and OMG34O4′ atoms (red)].

Because it is in a helical conformation, the anticodon (residues 34-36) is preformed for base pairing to the messenger RNA codon. The tight turn is believed to be a very common structural motif. This "U-turn" is stabilized by both base-backbone and backbonebackbone interactions: (1) a hydrogen bond from a U33 base proton to the A36 phosphate, (2) stacking of the A35 phosphate on the U33 base, and (3) a hydrogen bond from the U33 2′ OH to a base nitrogen from A35. Inversion of chain direction is achieved by a rotation of a single torsion angle, a (residue 34), to an unusual value. Finally, note the stacking of both the first and last loop nucleotides (residues 32 and 38) on the adjacent helix.

5.2.2 $tRNA^{Asp}$

As a result of synthetase binding, the other two anticodon loops shown in Figure elsewhere in this chapter have very different conformations, with additional tight turns. The $tRNA^{Asp}$ anticodon loop (upper left) could be considered a three-residue hairpin loop (residues 33-35) next to a two-residue bulge (residues 36 and 37).

Stacked as they are on the adjacent helix, residues 32 and 38 are in effect a helical extension. The reversal of the chain occurs gradually over residues 33-36, while two tight turns are found in the bulge (i.e., between residues 36 and 37, and between residues 37 and 38).

The tight turns in the bulge are stabilized by two hydrogen bonds:

the first is between the 2′-OH of residue 38 and the phosphate of residue 36; the second is between a base proton of residue 37 and a phosphate (not shown) from the anticodon stem. Note the close approach of phosphate groups in the bulge.

5.2.3 $tRNA^{Gln}$

Though superficially rather different in conformation, the $tRNA^{Gln}$ anticodon loop has in fact a conceptually similar structure. Unlike the $tRNA^{Asp}$ loop, synthetase binding to the $tRNA^{Gln}$ loop induces the formation of mismatched base pairs between residues 32 and 38, and 33 and 37.

These mismatches are stabilized by direct base-base as well as by water-mediated base-base hydrogen bonds. As in the tRNA^'P loop, a backbone-backbone hydrogen bond is found to stabilize a bulged-out conformation, but the location of this feature is different.

The hydrogen bond, found between the 2′-OH of residue 33 and the phosphate of residue 35, helps direct a large change in direction of the backbone. Additional tight turns are found among residues 35-37; indeed, residue 36 could be considered to be a bulge (note that the base of residue 35 is stacked on that of residue 37).

Thus this hairpin loop could be considered to be a two-residue loop (residues 34 and 35) next to a one-residue bulge (residue 36), much as the $tRNA^{Asp}$ anticodon loop could be considered to be a three-residue loop next to a two-residue bulge.

The overall structural similarity ol'these "anticodon bulge" conformations is probably due to the need of the synthetase to recognize the anticodon (residues 34—36), which encourages extrusion of these residues for case of recognition by the synthetase. Bulged residues are known to be recognized by proteins in several other systems.

The magnitude of the conformational changes induced by synthetase binding suggests that the anticodon loop may he quite flexible. This flexibility may facilitate adoption of conformations other than the "anticodon bulge" structure: for example, it is probable that the anticodon loop must adopt at least two rather different conformations during protein synthesis on the ribosome.

More generally, large loops not already involved in tertiary interactions may change conformation upon binding other RNA or protein molecules, and bulgelike conformations embedded within larger loops may be a common structural motif.

5.3 EMERGING PRINCIPLES OF NONHELICAL RNA STRUCTURE

Some of the generalizations from these and other structures may be summarized as follows.

Large loops are flexible. Binding to other RNA or protein molecules may result in large conformational changes. Bulged-out nucleotides embedded in a larger loop may be a common structural motif.

Small loops are more stable and less flexible. Smaller hairpin loops (e.g., the common four-nucleotide hairpin loops) are more thermodynamically stable and less flexible than larger hairpin loops. Because of their relative inflexibility, some small hairpin loops may be preformed for specific interactions with other molecules.

One or more mismatches stabilize the helix-loop interface Loop residues closest to a helix generally stack on the helix and may hydrogen-bond to form a mismatched base pair. In some cases, a pairing geometry very different from Watson-Crick geometry may be used to expose functional groups for tertiary interactions.

Base-backbone and backbone-backbone hydrogen bonds stabilize turns. Base-backbone hydrogen bonds are common at very tight turns; a common example is the U-turn motif. Hydrogen bonds between the 2′—OH proton and phosphate oxygens are often found to stabilize a bulged conformation; very often the bulged nucleotides are almost perpendicular to adjacent nonbulged nucleotides.

Base stacking is extensive but may be sacrificed in special cases. One or more bulged residues may be completely unstacked from a loop, often in order to engage in interactions with other RNA or protein residues.

5.4 RNA SECONDARY STRUCTRUES

RNA secondary structures can lie flat on a piece of paper with no overlapping strands. Unpaired regions between RNA helices include hairpin loops, bulge loops, internal loops, and multibranched loops or junctions. Secondary structure forms the core for the formation of the overall tertiary structure.

RNA secondary structures are important for RNA-RNA and RNA-protein interactions. Chemicals, ribonucleases, and site-directed mutagenesis probe these structures; comparative sequence analysis (phylogeny) and thermodynamic methods predict them.

5.4.1 Types of RNA Structure

Primary structure is the sequence of the bases, listed from the 5′

to the 3' end of the molecule. There are a few definitions of secondary structure, but the simplest is this: the structure lies flat on a sheet of paper with no strands overlapping.

Tertiary structure requires these secondary structures to interact out of the plane. Finally, quaternary structures are interactions between two separate molecules (e.g., RNA-RNA or RNA-protein).

5.5 TYPES OF SECONDARY STRUCTURE

Secondary structural elements come in five types: helices, bulges, internal loops, hairpin loops, and multibranched loops or junctions. Pseudoknots can be either secondary or tertiary interactions. The lower portion of Figure 1 annotates the secondary structure.

Helical or base-paired regions contribute most of the stability to RNA secondary structure through hydrogen bonding and base stacking. The Watson-Crick pairs, G·C and A·U, as well as some mismatches, such as G·U, stabilize a helix. Base stacking is such an important stabilizing effect that a single base stacking on the 3′ side of a helix (called a "dangling end") can add as much stability to the structure as a base pair.

Helices are on average six base pairs long in secondary structures of small subunit ribosomal RNA. A typical representation of helices in secondary structures is a ladder, where the rungs represent base pairs and the sides represent the sugar-phosphate backbone. Internal loops form when there are bases that cannot pair on both sides of a helix. Internal loops can be either symmetric (the same number of unpaired bases on each side of the helix) or asymmetric (a different number of unpaired bases on each side of a helix).

Two base internal loops are often called mismatches. Originally, studies of homopolymer loops (loops composed of one type of base) suggested that these regions would destabilize secondary structures; recent studies on common small internal loops reveal increased stability due to base stacking and non-Watson-Crick hydrogen bonding.

For example, the loop E region of *Xenopus laevis* 5S ribosomal RNA has an unusual structure with many nonWatson-Crick interactions. Internal loops are important sites of RNA-protein interaction in 5S rRNA, and proposed RNA-RNA tertiary and quaternary interactions in group I introns. Bulges are regions having unpaired bases on only one side of a helix; they can bend RNA backbones.

Bulges are important recognition sites for many regulatory and structural proteins. A single-bulged A is a common motif in 5S ribosomal

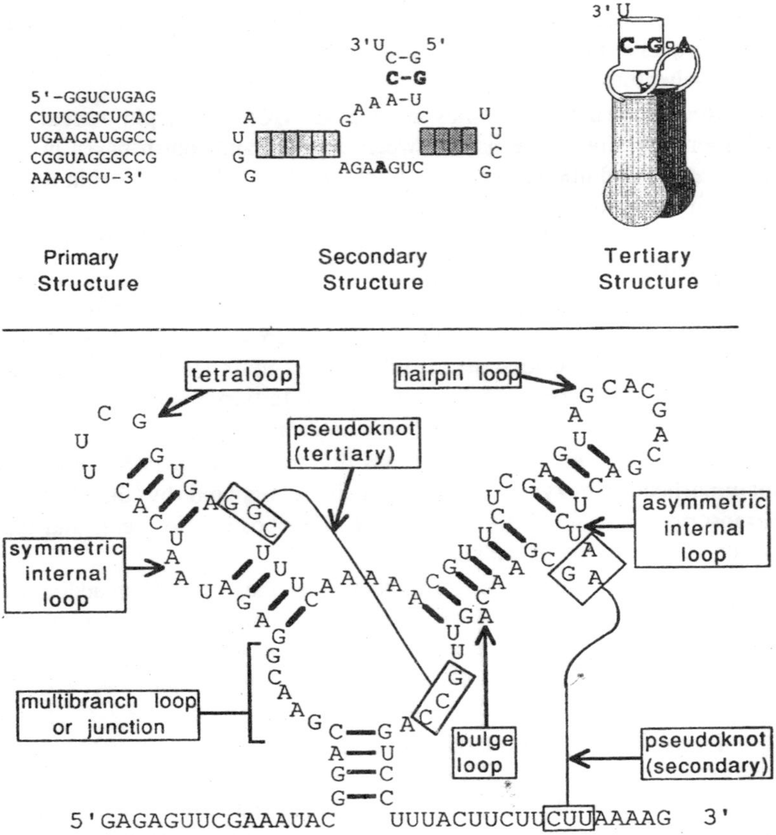

Figure 5.1: *Top:* The primary, secondary, and a possible tertiary structure of an RNA molecule. The tertiary structure drawing uses cylinders as helices and ribbons as single-stranded regions; it contains a tertiary interaction between a C·G base pair and an *A*. *Bottom: RNA* secondary structural elements.

RNA and is the guanosine binding site of group I intron ribozymes (a catalytic RNA). An unusual 2′—5′ linkage at a bulged A is a product of mRNA intron splicing by group II ribozymes, as well as the cellular splicing machinery (small nuclear ribonucleoprotein particles, or snRNPs).

Hairpin loops occur at the end of a helix when the sugarphosphate backbone folds back on itself to form an open loop; tracing the backbone reveals a hairpinlike structure. Comparisons of small subunit ribosomal RNA structures reveal an uneven distribution of hairpin loop sizes; four base loops are the most common.

Many four-base hairpin loops are either GGNRAC or CUNCGG

(loop bases underlined; N is any base, and R is either G or A), although other sequences are possible. These special sequence tetraloop hairpins often substitute for one another in ribosomal RNAs.

Recent structural studies on the CUUCGG hairpin reveal an additional strained base pair between the first and fourth bases in the loop, and an additional intraloop hydrogen bond. Larger hairpin loops can pair into complex structures involving nonWatson-Crick interactions. Hairpin loops are important for mRNA stability, RNA tertiary interactions, and protein binding sites.

Multibranched loops or junctions occur when three or more helices join to form a closed loop. The crystal structure of tRNAP" has a four-helix multibranched loop stabilized by helix-helix stacking as well as significant non-Watson-Crick secondary and tertiary interactions. These interactions probably stabilize other multibranched loops.

Three-helix multibranched loops are found in 5S rRNA and the hammerhead ribozyme; the former is recognized by the TFIIIA (transcription factor A for polymerase III) ribosomal protein, and the latter is the site of phosphodiester cleavage.

Pseudoknots are special RNA interactions; they result from additional pairing in loop regions of the aforementioned secondary structures. Pseudoknots can form either secondary structures (such as a local folding back on a hairpin structure) or tertiary structures spanning several helical regions.

Pseudoknots are important structural features in the tRNA-like ends of viroids, 16S rRNA of *Escherichia coli,* and the catalytic core of the group 1 introns. They are important recognition sites for regulatory proteins, and they cause translational frameshifting in many viruses.

5.6 PREDICTING SECONDARY STRUCTURE

RNA secondary structure is predicted by comparative sequence analysis (phylogeny) or thermodynamics. Comparative sequence analysis finds a common folding pattern of two or more RNAs having a similar function. The existence of compensatory base pairs is proof of helical regions; for example, an A·U pair in one sequence may replace a C·G pair in another sequence.

This method cannot predict structure in conserved sequence regions, since there are no compensatory base pairs. Nevertheless, it is the most accurate way of predicting Watson-Crick pairing regions. Thermodynamic predictions assume that the solution structure has the largest equilibrium constant of formation, or, in other words, the most negative Gibbs free energy (AG°). RNA secondary structure stability is based on the strong local interactions of base stacking and hydrogen bonding; these local interactions are measured for small oligoribonucleotides and used to predict the structure of large RNAs.

Thermodynamic predictions succeed if there is a measured OG° for each interaction. Even though the stacking and hydrogen bonding schemes in loop regions are not completely understood, a recent thermodynamic model predicts about 70% of the helices of phylogenetically determined secondary structures.

Secondary structure prediction programs based on thermodymanics let the user include phylogenetic data as well as constraints from chemical modification and ribonuclease digestion.

5.7 PROBING SECONDARY STRUCTRUE

Ribonucleases, chemical modification and cleavage, cross-links, and mutagenesis are four direct probes of RNA secondary structure.

Other methods, such as thermal denaturation, kinetics, and nuclear magnetic resonance spectroscopy, are not discussed here.

Experiments must be conducted such that ribonucleases and chemicals modify only once per RNA molecule, since modification or cleavage may alter the secondary structure. Ribonucleases are the simplest RNA structure probes; some enzymes cleave helical regions (such as ribonuclease V_1), while others (T_1, T_2, S_1, and CL_3) attack only single-strand regions.

Base-modifying reagents, like kethoxal or dimethyl sulfate, are important secondary structure probes, whereas chemicals that modify phosphates (e.g., ethylnitrosourea) or cleave the sugar-phosphate backbone [e.g., iron-EDTA (ethylenediaminetetraacetic acid)] are more important as tertiary and quaternary structure probes. Cross-links (unusual covalent bonds) form between two strands of a pairing region with ultraviolet light or special reagents, such as psoralen.

Detection of RNA cleavage or modification is monitored directly using end-labeled RNA or indirectly using reverse transcription. After cleaving 5′- or 3′-^{32}P-labeled RNA with ribonucleases or chemicals, the different-sized strands can be separated by denaturing polyacrylamide gel electrophoresis.

Each radioactive band on the gel represents a different cleavage site; the longer the fragment, the further the strand cleavage site from the labeled end. Reverse transcription can detect strand scission, but can also detect base modifications and cross-links. Reverse transcriptase synthesizes a complementary DNA strand from the RNA template starting at a labeled DNA primer; it will stop at modified bases and at the end of the RNA.

A combination of chemical modification and reverse transcription was used to confirm the secondary structure of the *E. coli* 16S ribosomal RNA, which was originally determined by phylogenetic analysis. Mutational analysis creates an artificial phylogeny using mutagens or site-directed mutagenesis.

Mutagens induce nonspecific changes in an organism's genome; the sequences of viable strains can show compensatory mutations. In contrast, site-directed mutagenesis creates specific mutations in a cloned sequence.

The effects of a single base mutation in the helix and of the compensatory base change are observed by a functional assay, such as protein binding or catalytic activity. Mutational analysis is an important tool to determine secondary structure in conserved sequence regions, such as the catalytic core of group I introns.

5.8 RNA REPLICATION

As more has been learned about the process of nucleotide replication it has become evident that DNA replication, RNA replication, and DNA transcription are simply variations on a common theme, that of storing information and then reading out that information as the cell grows and divides.

All these processes depend on the formation of a polynucleotide strand, which serves as a template for the production of a complementary new strand of polynucleotide aligned by base pairing of adenine (A) to thymine (T) and guanine (G) to cytidine (C). The replication processes are carried out with great accuracy, with less than one error per thousand base pairs formed.

5.8.1 Reverse Transcriptase

The central dogma of molecular biology, unchallenged for several decades, was that genetic information always flowed from DNA to RNA to protein. A seminal exception to this rule was discovered in 1964 by Howard Temin. Temin's experiments showed that infection of cells by certain cancer-causing viruses is blocked by inhibitors of DNA synthesis and by inhibitors of DNA transcription.

These experiments indicated that the synthesis and transcription of DNA are required for the multiplication of RNA tumor viruses and that in these systems information flows in the opposite direction, from RNA to DNA. Temin suggested that a DNA provirus was essential in the replication of RNA tumor viruses. In these organisms an enzyme was required that would synthesize DNA using RNA as a template.

In 1970 Temin and Baltimore identified just such a DNA polymerase, one that uses RNA as a template in the synthesis of DNA. This RNA-dependent DNA polymerase, now termed "reverse transcriptase," has been shown to be present in all RNA tumor viruses. Other RNA viruses that do not form tumors replicate themselves directly, without using DNA as an intermediate.

5.9 STRUCTURE OF DNA AND RNA

Both DNA and RNA are long strands of nucleotides joined together by 3′,5′-phosphodiester bonds. Each nucleotide is composed of a purine or pyrimidine base, a sugar (ribose in RNA and deoxyribose in DNA), and orthophosphate. DNA is composed of four kinds of nucleotide: adenylate (A), guanylate (G), cytidylate (C), and thymidylate (T). RNA contains uridylate (U) instead of T.

Uridylate has the same binding properties as thymidylate and base-pairs with adenylate. RNA is usually a single-stranded molecule. Although RNA has only a single-stranded polynucleotide chain, many of the sequences in RNA are mutually complementary, and these sequences can undergo intrachain base pairing to form stem-loop (hairpin) structures.

5.9.1 Structure of DNA

The DNA molecule consists of a very long double helix. The two chains that make up the helix extend in opposite directions and are paired by the Chargaff rules: a nitrogenous base in one chain is paired to a base in the other chain such that an A in one chain always pairs to a T in the other chain and a G pairs with a C.

The specific base pairing is determined by the nature of the hydrogen bonds joining the two bases: two bonds join A and T and three hydrogen bonds join G and C. To fit into the double helix, one large base, A or G, must pair with a smaller base, C or T. An A·G pair would be too large and a C·T pair would be too small to fit in the space available in the double helix.

In DNA replication the two chains separate momentarily and, with the aid of a number of proteins (some of which are enzymes), a new chain is formed by base pairing to each of the original chains. The product is two DNA molecules, each composed of a double helix.

5.9.2 Information Transfer: DNA to DNA

The Watson-Crick model of DNA suggested a mechanism by which the information in DNA could be copied precisely. Since nucleotides pair with each other in a complementary fashion, A to T and G to C, each of the nucleotide strands in the DNA molecule could serve as a template for the synthesis of its opposite strand.

When the hydrogen bonds joining the two strands are broken, the two strands can separate. Each chain can then pair with complementary nucleotides to form the corresponding strand. This results in two DNA double helices, both of which are identical to the original one.

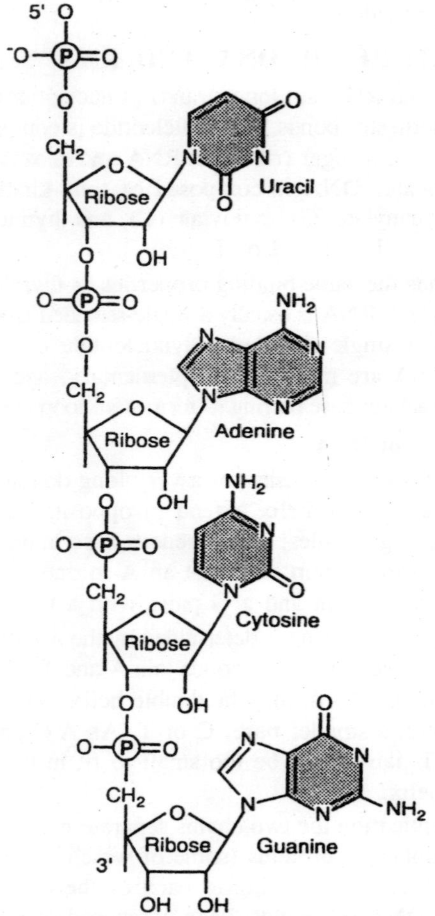

Figure 5.2: A portion of an RNA strand, with ribonucleoside phosphates linked by 5'-3' bonds, like the deoxyribonucleosides in DNA.

Each consists of one original strand from the parent molecule and one newly synthesized complementary strand. The replication of DNA is termed "semiconservative" because each of the original strands is conserved in one of the daughter double strands and constitutes half of the daughter helix.

The two strands of DNA in the double helix are physically intertwined; they must be separated for replication to proceed. The separation of the strands has proved to be a complex process. Replication can occur in these very long DNA molecules only if the strain of the unwinding strands is relieved.

The unwinding is catalyzed by DNA helicases, enzymes that move along the helix, unwinding the strands as they go. The separated strands are bound by helix-destabilizing proteins that bind to single-stranded DNA, preventing the reestablishment of the double helix until each strand has been copied.

The DNA polymerases that link together the nucleotide subunits in the replication of DNA add nucleotides to the 3′ end of a polynucleotide strand that is paired to the strand being copied. The substrates for the DNA polymerases are dexoyribonucleoside triphosphates. As the nucleotides are joined, two of the phosphates are removed, providing the energy to drive the synthetic reaction.

A new polynucleotide chain is elongated by the addition of the 5′ phosphate group of the next nucleotide subunit to the 3′ hydroxyl sugar at the end of the growing strand. Thus, DNA synthesis proceeds in a 5′ → 3′ direction. DNApolymerases can catalyze the addition of nucleotides only at the 3' end of an existing DNA strand.

This leads to the question of how the synthesis of DNA can be initiated when the two strands are separated. This is accomplished by utilizing a short piece of an RNA primer that is synthesized by an aggregate of proteins called a primosome.

The RNA primer pairs with a single-stranded DNA template at the point of initiation of replication. DNA polymerase can then synthesize the new chain by adding nucleotides to the 3' end of the RNA primer. When DNA synthesis has proceeded to an appropriate extent, the RNA primer is degraded and removed by specific enzymes and that space is filled in by DNA polymerase.

5.9.3 DNA Replication: Leading and Lagging Strands

Although the complementary DNA strands extend in opposite directions, DNA synthesis can proceed only in the 5'—3' direction. The strand being copied is read in a 3′→5′ direction. DNA replication is begun at specific sites on the DNA molecule, called origins of replication, and both strands are replicated at the same time.

One, the leading strand, is formed continuously after the process has been initiated. The complementary (lagging) strand is synthesized in short pieces, which are subsequently joined to make a complete DNA chain. DNA is synthesized beginning at a Yshaped structure called a replication fork.

The lagging strand is synthesized in the form of relatively short fragments (Okazaki fragments), each of which is initiated by a separate

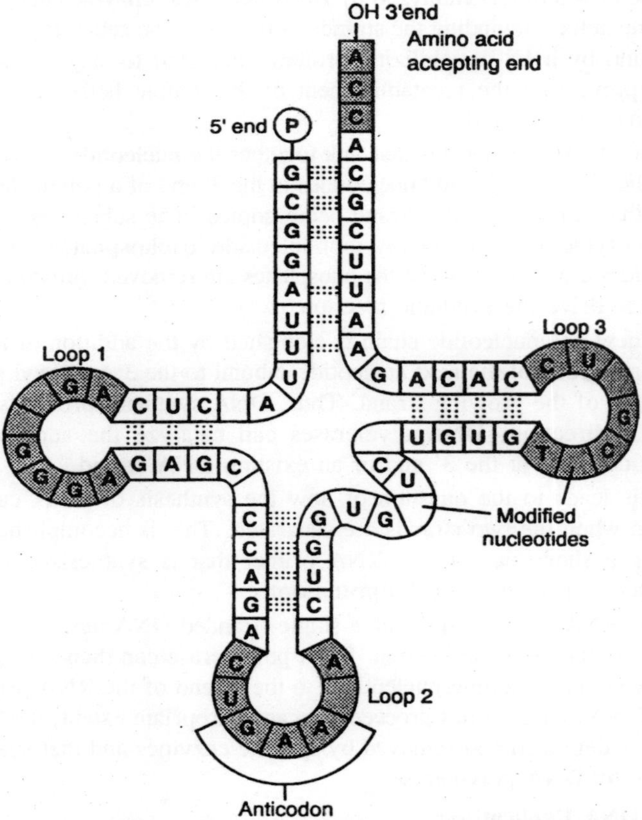

Figure 5.3: The structure of a transfer RNA molecule, showing the many sites of complementary base pairing (dots indicate hydrogen bonding between A and U and between C and G). Similar complementary base pairing occurs frequently in messenger RNA as well.

primer and is then extended toward the 5′ end of the previously synthesized fragment by a DNA polymerase. The complex DNA polymerases serve several functions.

As the growing fragment approaches the fragment synthesized previously, one part of the polymerase degrades the RNA primer, permitting other polymerases to fill in the gap between the two fragments. These are then linked by DNA ligase, which joins the 3' end of one fragment to the 5′ end of another fragment by a phosphodiester bond.

Since similar codes are used in both DNA and RNA, in certain retroviruses, such as the AIDS virus, information is transferred from RNA, which is the genetic material of the retroviruses, to DNA, which becomes an intermediate in the synthesis of the next generation of RNA. Information is transferred directly from RNA to RNA in certain RNA viruses such as the influenza virus or the poliomyelitis virus.

5.9.4 Information Transfer: DNA → RNa

The transcription of DNA to form RNA and the replication of RNA are processes that are basically similar to that of DNA replication discussed in other section of this book. In all these processes one polynucleotide strand serves as a template, and specific polymerases use specific nucleoside triphosphates to add specific nucleotides to the chain. The particular base added in sequence is determined by the complementary pairing of the bases (determined by the hydrogen bonds between the bases) in the initial and forming strands.

Information present in the sequence of nucleotides in one of the DNA strands, the minus strand, is copied, yielding a plus strand of messenger RNA. This transcription process is very similar to that by which DNA is replicated: the plus mRNA strand is formed by complementary base pairing with the minus strand of DNA.

Information that has been transcribed into mRNA is used to determine the amino acid sequence in the protein.

It is clear that this process involves the conversion of the nucleic acid code of the messenger RNA into an amino acid code of the protein. Hence it is termed a process of translation. The information present in the mRNA is in the form of a genetic code composed of three bases that constitute a codon, the basic unit of genetic information.

Each group of three bases in mRNA specifies.one of the amino acids in the protein. The codon that specifies the amino acid tryptophan is UGG. Any of six different codons (CGU, CGC, CGA, CGG, AGA, AGG) may specify the amino acid arginine.

The information contained in the codons of mRNA is translated by a very complex process that occurs within the ribosomes in eukaryotic cells. The recognition and decoding of the codons in mRNA is accomplished by transfer RNAs. The anticodons of the tRNA are composed of triplet nucleotides, which recognize a specific codon in mRNA by complementary base pairing.

The amino acid specified by the anticodon is attached to the other end of the tRNA molecule. Protein synthesis occurs on the ribosomes, complex suhcellular organelles composed of two kinds of subunit. Each contains one or more kinds of ribosomal RNA (rRNA) and a considerable number of proteins.

The ribosomes are attached to the 5′ end of the mRNA. The ribosome travels along the mRNA, joining the codon of the mRNA with the anticodon of the tRNA by base pairing. In this way the amino acids are lined up in the proper sequence. The amino acids are then joined enzymatically to form a long polypeptide chain.

5.9.5 Sense and Antisense Strands

Only one of the two strands in DNA contains the correct coding sequence for protein synthesis. This, the minus or sense strand, undergoes transcription to form mRNA. The complementary DNA strand is termed plus, or antisense. The RNA polymerase that catalyzes transcription recognizes a specific promoter sequence of bases at the 5′ end.

Messenger RNA is synthesized by the addition of nucleotides, one at a time, to the 3' end of the growing molecule. The initiation of RNA replication does not require a primer. The 5′ end of a new mRNA chain begins with a nucleoside triphosphate. The promoter sequences of different genes may differ and this can determine which genes will be transcribed at any given time.

Three of the codons in mRNA (UGA, UAA, UAG) serve as "stop" signals for the RNA polymerase and bring about the termination of transcription. The mRNA molecules Of eukaryotic organisms undergo posttranscriptional modifications and processing.

A molecule of 7-methylguanylate, an unusual nucleotide, is added as a cap at the 5′ end of the mRNA chain. This alteration may be the basis of the greater stability of eukaryotic mRNAs, which have half-lives as long as 24 hours, whereas the half-life of prokaryotic mRNA is about 15 minutes. A long tail of polyadenylic acid, 100-200 adenine nucleotides, is joined to the 3′ end of the mRNA.

5.9.6 Introns and Exons

A third step in the modification of eukaryotic mRNA involves cutting and splicing the mRNA at specific sites. Interrupted coding sequences are present in eukaryotic DNA; these may be quite long, but they do not code for the amino acids present in the final protein product. These noncoding regions within the gene are called intervening sequences or introns, as opposed to exons (which are expressed sequences and part of the protein-coding sequence).

The number of introns present in a gene may vary widely. The β-globin gene, which produces one of the components of hemoglobin, contains two introns. The ovalbumin gene, which codes for one of the proteins in egg white, contains seven introns, and the gene for another egg-white protein, conalbumin, contains sixteen.

The combined lengths of the introns may be considerably greater than the combined exon sequences. The ovalbumin gene contains about 7700 base pairs, whereas the sum of its coding sequences (exons) is only 1859 base pairs.

The transcription of a gene containing both introns and exons yields a large RNA transcript termed heterogeneous nuclear RNA. For the hnRNA to be converted into a functional mRNA, the introns must be deleted and the exons must be spliced together to form a continuous message that codes for a specific protein.

The splicing reactions are mediated by special base sequences within and to either side of the introns. The splicing reactions may be catalyzed by small nuclear ribonucleoprotein complexes, which bind to the introns and catalyze the cleavage and splicing reactions. The RNA within the intron may have the ability to splice itself without the help of protein catalysts; these nucleic acids act as enzymes.

This new class of biological catalysts has been termed ribozymes. The final product, functional mRNA, has had its introns removed and the exons spliced together. It has a 7-methylguanylate cap at the 5′ end and a polyadenylated tail at the 3′ end. It is then ready to pass out of the nucleus into the cytoplasm and, when attached to a ribosome, to serve as the blueprint for the synthesis of a protein with its specific sequence of amino acids.

5.10 VIRAL DNA AND RNA

Much of what we know about genetic machinery in general and about replication and transcription in particular has been derived from studies of viruses. The viruses that infect animal cells exhibit a tremendous variety of shapes, sizes, and genetic strategies. Viruses are classified according to the sequence of reactions by which the mRNA is produced.

In this classification scheme, a viral mRNA is termed a plus strand and its complementary sequence. which cannot function as a messenger RNA, is termed a minus strand.

A strand of DNA complementary to the viral mRNA is termed a minus strand. Production of a plus strand of mRNA requires that a minus strand of RNA or DNA be used as the template. This arrangement permits the characterization of six kinds of virus that infect animal cells.

In each of these, the nucleic acid of the virion (the infective particle) ultimately becomes the mRNA of the daughter virus. The viruses that infect animal cells have genomes composed of DNA or RNA. In some the nucleic acid is single-stranded and in others the nucleic acid is double-stranded.

The adenoviruses and the SV 40 virus contain double-stranded DNA. The viral DNA enters the nucleus of the host cell. The enzymes that in the uninfected state are responsible for producing cellular mRNAs are diverted to producing viral mRNA.

The pox viruses are large viruses, which have their own enzymes for synthesizing mRNA; they undergo replication in the cytoplasm of the host cell. Parvoviruses are simple viruses containing a single strand of DNA. Some parvoviruses enclose within their capsules both plus and minus strands of DNA, but the strands are present in separate virions.

Other viruses enclose only a minus strand within the capsid, and this is copied within the cell into double-stranded DNA, which is then copied to yield messenger RNA.

5.11 INFORMATION TRANSFER RNA TO RNA

Four additional kinds of virus contain RNA rather than DNA genomes. A wide range of animals, including human beings, may be infected by viruses belonging to each of these classes. Viruses with a genome of double-stranded RNA utilize the minus RNA strand as a template for the synthesis of a plus strand of mRNA.

The virions of these viruses have segmented genomes containing 8-12 double-stranded RNA fragments, each of which codes for a specific polypeptide. These viruses contain a complete set of enzymes for the synthesis of mRNA.

In viruses with a genome composed of a single plus strand of RNA. the viral genomic RNA is identical to the mRNA and the genomic RNA is infectious by itself. The mRNA is copied into a minus strand of RNA, which then produces more plus strands. The RNA in the virion of poliomyelitis virus serves as the mRNA to encode all the viral proteins. The proteins are initially synthesized as a single, very long polypeptide chain, which is subsequently cleaved to yield the several functional proteins. The mRNA of these viruses is the same length as the genomic RNA.

The virions of toga viruses are surrounded by a lipid envelope. These viruses synthesize at least two forms of tnRNA within the host cell; one is the same length as the RNA in the virion, whereas the other corresponds to the third of the virion RNA at the 3' end. This is one of a group of rare insect-borne viruses that cause encephalitis in human beings.

A further class of viruses contain a single minus strand of RNA. The RNA in the virion has a base sequence complementary to that of the mRNA. Thus, the virion contains a template for making mRNA, but that template does not itself encode proteins.

Viruses with a single strand of RNA in their genome have in the virion a virus-specific polymerase that synthesizes several different mRNAs from different parts of the single RNA template strand. Each viral mRNA encodes one protein. The influenza virus also has a segmented genome.

Each segment is a template for the synthesis of a single mRNA. The minus strands of these viruses alone, in the absence of the virus-specific polymerise, are not infectious. The RNA polymerase of the influenza virus initiates the transcription of each mRNA by a unique mechanism.

The polymerase begins the synthesis of mRNA by borrowing 12-15 nucleotides from the 5′ end of the cellular mRNA or the mRNA precursor in the nucleus. This oligonucleotide serves as a primer for the replication of RNA catalyzed by the viral RNA polymerase. The individual mRNAs made by these viruses generally encode single proteins, but some of the mRNAs can be read in different frames to produce two different proteins.

Since some viruses contain RNA but not DNA, in these organisms RNA must be the viral genetic component. This theory was confirmed by the finding that purified RNA preparations from tobacco mosaic virus were infectious even in the absence of any TMV protein. Many other viruses have been shown to contain RNA but not DNA. Thus the potential of RNA for carrying genetic information and undergoing replication cannot be denied.

In most viruses the DNA or RNA codes not only for the coding proteins of the virus but also for the enzymes that are required to replicate the viral nucleic acid. The smaller DNA viruses such as the simian SV-40 virus, and the very small bacteriophage X174, contain very much less genetic information and must rely to a much greater extent on enzymes from the host cells to carry out the synthesis of proteins and DNA.

These viruses do contain coding for enzymes that initiate their own DNA synthesis selectively. For a virus to be successful when it invades a cell, it must override the cellular control signals that otherwise would prevent the viral DNA from doubling more than once in each cell cycle. The discoveries in the 1950s and 1960s that many of the smaller DNA viruses contain only single-stranded DNA can now be appreciated, although at the time it seemed that viruses might have a fundamentally different mode of replication.

These DNA viruses begin replication by producing a double-stranded intermediate. This finding reinforced the concept that all DNA replication, all transcription of DNA to RNA, and all RNA replication involve the formation of complementary strands of polynucleotides.

5.12 REPLICASES

The replication of RNA and the multiplication of RNA viruses also involve the formation of complementary chains of nucleotides joined by specific hydrogen bonds between specific purines and pyrimidines.

RNA replication is mediated by specific RNA-dependent RNA polymerases (called replicases), which are coded for in the viral RNA chromosome. Some of the RNA viruses, such as poliovirus, have a genome composed of a single-stranded RNA polynucleotide chain. Other viruses, such as the reoviruses, have a double-helical RNA viral chromosome.

5.13 RETROVIRUSES: INFORMATION TRANSFER FROM RNA TO DNA

The genomic RNA of retroviruses is a single plus strand, which serves as a template that directs the formation of a DNA molecule. The process is catalyzed by an RNA-dependent DNA polymerase. The DNA produced in turn serves as the template for the production of mRNA.

First the virion RNA is copied into a single strand of DNA, which then forms a complementary second DNA strand. The double-stranded DNA is integrated into the chromosomal DNA of the host cell. Finally the integrated DNA is transcribed by the cell's own metabolic machinery into RNA that either acts as a viral mRNA or becomes enclosed in a virus. This completes the retrovirus cycle. If the retrovirus contains oncogenes (cancer genes), the cell it infects may be transformed into a tumor cell.

5.14 EVOLUTION OF COMPLEMENTARY BASE PAIRING

The evolution of the process of complementary base pairing was a key step in the evolution of life, for it permitted the development of mechanisms for the transfer and storage of genetic information. Most scientists now believe that RNA is the more ancient means of transferring genetic information and that RNA replication preceded DNA replication in the course of evolution.

Thus, RNA replication may have been the evolutionary precursor of DNA replication. The replication of RNA and DNA and the transcription of DNA to RNA and the translation of RNA to form proteins all involve this important basic principle of complementary base pairing, determined by specific hydrogen bonds between the base pairs A·T and G·C.

In protein synthesis the lining up of the amino acids is determined by the complementary base pairing of codon and anticodon. The course of evolution of nucleic acids may have been as follows:

		Virus with single-stranded DNA	→	Virus with double-stranded DNA
Virus with single-stranded RNA	→	Virus with double-stranded RNA		
	↘	Virus with RNA/DNA hybrid		

6

Translocation of RNA to Protein

Proteins are essential to the structure and function of living cells. The assembly of polypeptide chains from amino acids and their subsequent modifications, leading to the final three-dimensional protein structure, are exceptionally complex processes; many components are involved and much of the cell's energy is utilized.

The linear amino acid sequence of a protein is encoded within the gene as a linear deoxyribonucleotide sequence. Early steps in the biosynthesis of a protein include transcription of the gene and appropriate processing of the transcript leading to the production of mature mRNA.

We describe the mechanisms involved in translating mRNA to produce a polypeptide chain which has the amino acid sequence specified by the gene. A gene or cistron is defined as the region of DNA that is transcribed into functional RNA.

The transcript functions either as such (e.g., tRNA, rRNA) or as a messenger (mRNA), which codes for a single polypeptide chain in the translation process. A polynucleotide such as RNA is an asymmetrical polymer that is assembled from nucleoside triphosphates by a stepwise mechanism linking the 3′ position of one nucleotide by a phosphate bridge to the 5′ position of the adjacent nucleotide.

In the finished polynucleotide chain, the first nucleotide residue has a 5′ position that is not linked to another nucleotide, whereas the last nucleotide has an unlinked 3′ position. Thus, polynucleotide synthesis proceeds from the 5′ to the 3′ terminus and the polymer is said to have a 5′-to-3′ polarity. Usually, linear RNA sequences are written with the 5′ terminus on the left and the 3' terminus on the right.

The polypeptide chains of proteins are also asymmetrical polymers in which the amino acid residues are linked by peptide bonds between their α-amino and carboxyl groups, leaving a free α-amino group at one end (the amino terminus) of the polymer and a free α-carboxyl group at the opposite end (the carboxyl terminus).

The significance of the polarity of RNA and proteins will become evident when the process of protein biosynthesis is explained. The genetic information stored in DNA is not usable directly for making proteins but must be copied first into mRNA by an enzymatic transcription of segments of DNA containing the genes.

Messenger RNA serves as the template for protein synthesis; that is, the linear nucleotide sequence of the mRNA dictates the amino acid sequence of the polypeptide encoded originally by the gene.

Conventionally, gene and mRNA nucleotide sequences are written in the 5′-to-3′ direction, which corresponds to the direction in which mRNA is decoded during polypeptide synthesis: the mRNA is read in the 5′-to-3′ direction, and the polypeptide is synthesized from the amino- toward the carboxyl terminus.

The mechanism for translating RNA into protein is complex, and the cell devotes considerable resources to the translational machinery. The components include 20 different amino acids, transfer RNAs, aminoacyl-tRNA synthetases, ribosomes, and a number of protein factors that cycle on and off the ribosomes and facilitate various steps in initiation of translation, elongation of the nascent polypeptide chain, and termination of synthesis with release of the completed polypeptide from the ribosome. The process depends on a supply of energy, which is provided by ATP and GTP. The rate of protein synthesis is approximately 0.15 second per peptide bond at 37°C.

6.1 MRNA STRUCTURE AND THE GENETIC CODE

6.1.1 Structure

The sequence information of a gene is copied (transcribed) into the nucleotide sequence of RNA using one strand of DNA (called the coding strand) as the template. The primary transcript is a single strand of RNA, which is a faithful copy of the *other* strand of DNA (the non-coding strand), with substitution of U residues in place of T residues found in DNA.

Sometimes, the primary transcript is altered, as described later, before it functions as mRNA; in these cases the original unmodified transcript is the precursor or pre-mRNA. Usually, mRNAs have nontran-

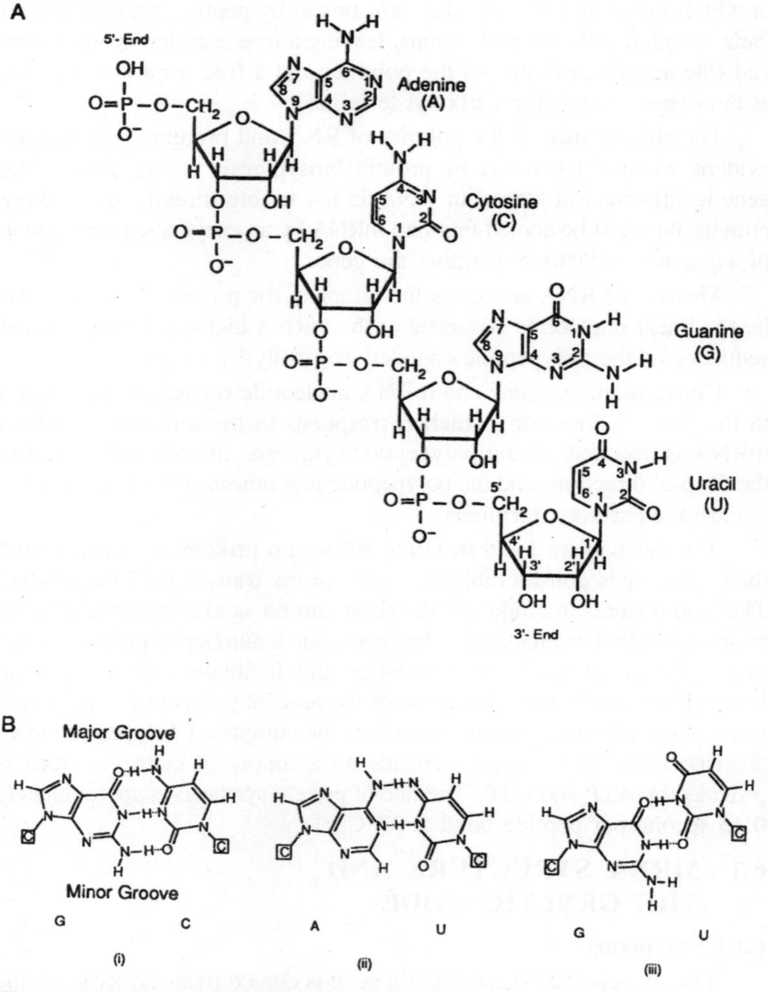

Figure 6.1: Structural elements of ribonucleic acids.

slated sequences at the 5′ and 3′ ends in addition to the coding domain. These noncoding sequences sometimes affect the efficiency of translation and the stability of mRNA. Figure elsewhere in this chapter shows the structures of typical mRNAs.

The decoding process involves base pairing between three bases (i.e., a codon) in the mRNA and the threebase anticodon of a transfer RNA. In a separate reaction, each tRNA is first linked to a particular

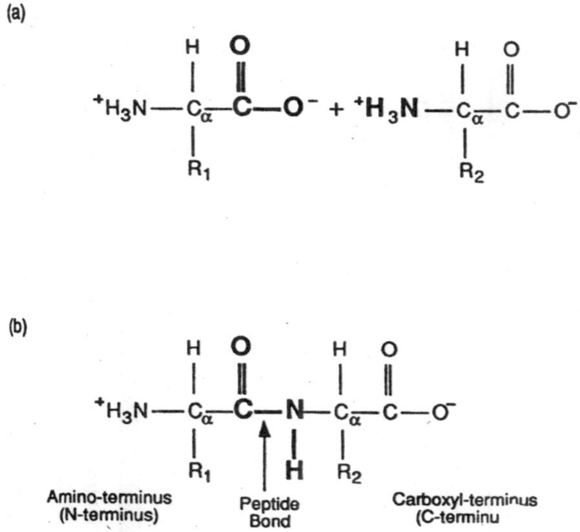

Figure 6.2: Structure of the peptide bond.

amino acid, and thus the pairing of mRNA with tRNA determines the sequence of amino acids in the resulting protein.

6.1.1.1 Prokaryotic mRNA

In organisms that do not have a nucleus (prokaryotes), pre-mRNA usually undergoes little or no modification, with the result that pre-mRNA and mRNA are very similar if not identical. Since premRNA is collinear with DNA, DNA and proteins are usually collinear in these organisms.

Gene expression in prokaryotes usually involves the cotranscription of several adjacent genes, and translation of mRNA sequences into polypeptides may begin at the 5′ end of mRNA while transcription is still in progress at the 3′ end.

6.1.1.2 Eukaryotic mRNA

In cells with a nucleus (eukaryotes), the genetic information is stored mainly in the nucleus and to a minor degree in some organelles (mitochondria and chloroplasts). The description that follows pertains only to nuclear genes.

Eukaryotic genes are more complicated than prokaryotic genes because the coding region in the former is often discontinuous: the coding sequences or exons are interrupted by intervening sequences (introns). Thus, genes and proteins are usually not collinear in eukaryotes. In the nucleus, a complicated set of splicing reactions removes all the

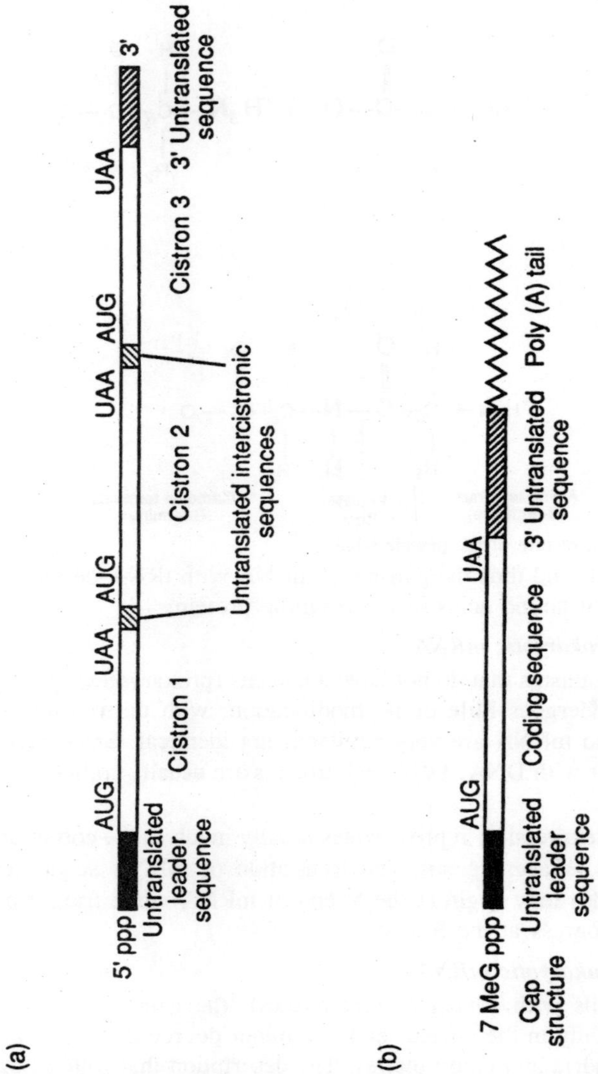

Figure 6.3: Structure of typical mRNAs.

introns and fuses the exons into a continuous coding sequence. Other processing steps involve adding a "cap" to the 5′ end of the mRNA adding a polyadenylated tail to the 3′ end.

After completion of these nuclear maturation steps the mRNA is transported to the cytoplasm, where it is translated. As with prokaryotic mRNA, the coding region is flanked by 5′ and 3′ nontranslated sequences.

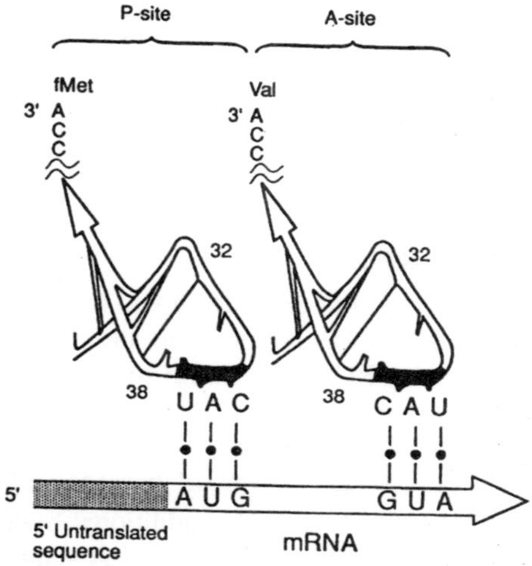

Figure 6.4: Schematic illustration of base pairing between a codon and its auticodon.

6.1.2 The Genetic Code

With a few minor exceptions (e.g., in mitochondria), the genetic code is universal. It is triplet, comma-less, and nonoverlapping, and, as a consequence, a nucleotide sequence has three possible reading frames. Since mRNA is normally translated into a unique polypeptide, an essential step in the translation process is the selection of the appropriate reading frame.

This is achieved by starting translation at the initiation codon, usually AUG or less frequently GUG, which ensures that the following codons are read in phase within the required reading frame. Of the 64 theoretically possible triplets in the genetic code, 61 correspond to the 20 genetically encoded amino acids found in all, or nearly all, proteins.

When GUG is used as the initiator codon, it interacts with the anticodon of initiator Met-t RNA;", but elsewhere it codes for valine. All other codons specify only one amino acid, but many amino acids are specified by two or more (up to six) codons; that is, the code is unambiguous but degenerate.

The remaining three codons usually signify termination of synthesis and release of the finished polypeptide chain. The decoding process involves antiparallel base pairing between the three bases of mRNA codons and the complementary anticodons of tRNA during peptide bond formation.

The first and middle bases of the codon form conventional base pairs with the third and middle bases of the anticodon, respectively, but the third base of the codon pairs with the first base of the anticodon by a less stringent interaction. This so-called wobble considerably reduces the number of tRNA species required to decode the 61 sense codons.

Thus, the protein synthesis system in the cytosol of eukaryotes contains only a few more than 40 different tRNAs, and in mitochondrial protein synthesis 22 to 24 tRNA species are sufficient.

6.2 TRANSFER RNA

By relating individual codons of mRNA to the cognate amino acids, tRNA functions as a key bilingual intermediate in the translation of the genetic code. All transfer RNAs are single-stranded molecules about 80 nucleotides long, with a common 3'-terminal CCA sequence. The secondary structure of tRNA is usually presented in two dimensions as a clover leaf, to highlight the regions of base pairing.

X-Ray crystallography reveals that additional hydrogen bonds give rise to an L-shaped tertiary structure. The CCA sequence carrying the amino acid is located distal to the anticodon. During translation, the three nucleotides of the anticodon interact, through the formation of antiparallel base pairs, with the three nucleotides of the codon.

The attachment of amino acids to tRNA' involves the formation of an ester bond between the carboxyl group of the amino acid and the 3'-hydroxyl group of the terminal adenosine. It requires specific enzymes, the aminoacyl-tRNA synthetases.

There are 20 different synthetases, each specific for one of the 20 amino acids, and each enzyme recognizes something unique in the structure of its cognate tRNA. The structural determinants that ensure the accuracy of this charging reaction vary among the different tRNAs. The anticodon may play a part, but sometimes even a single base elsewhere is sufficient to determine the specificity of the tRNAsynthetase interaction.

The accuracy of the synthetase reactions is important because the structure of the amino acid plays no part in its subsequent selection by the mRNA for peptide bond synthesis.

The energy for the reaction is provided by ATP and stored in the ester bond of the aminoacyl-tRNA to be used subsequently for peptide bond synthesis.

6.3: RIBOSOME STRUCTRUE AND FUNCTION IN TRANSLATION

6.3.1 Ribosome Structure

Ribosomes are high molecular weight complexes of RNA (rRNA)

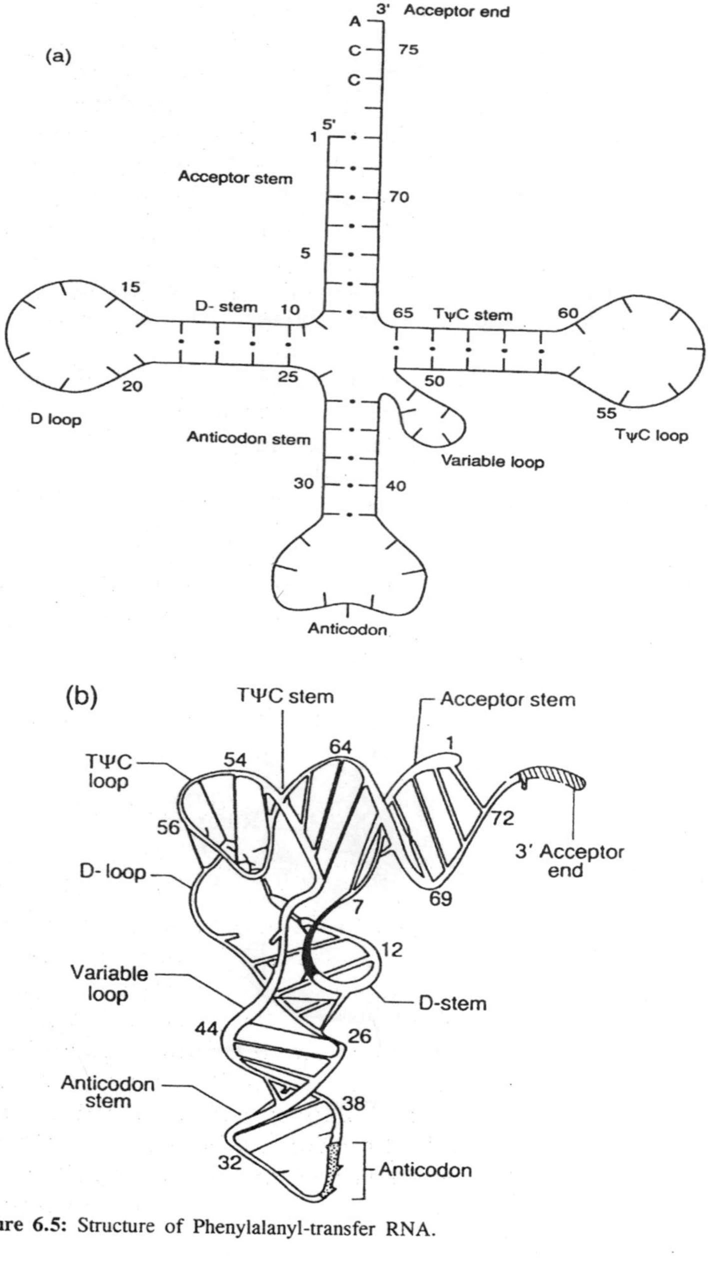

Figure 6.5: Structure of Phenylalanyl-transfer RNA.

and proteins. Ribosomes from various sources (prokaryotes, eukaryotic cytoplasm, mitochondria, and chloroplasts) vary in size from 20 to 30 nm in diameter, but all are composed of a large and a small subparticle or subunit and perform similar functions in protein synthesis.

The small subunit comprises a single rRNA of 0.3 to 0.7 $\times$ 10^6 Da and single copies of 20 to 30 unique proteins. It has a major function in binding initiator tRNA and mRNA in the initiation of protein synthesis and in decoding the genetic message.

The large subunit comprises a high molecular weight rRNA (0.6-1.7 $\times$ 10^6 Da) and often one or two smaller rRNAs (0.03-0.05 $\times$ 10^6 Da) and 30 to 50 different proteins, present, with one exception, as single copies. The large subunit binds aminoacyltRNA at the A site, peptidyl-tRNA at the P site, and uncharged tRNA at the E site.

The large subunit contains the peptidyl transferase activity, which may reside in the rRNA molecule itself rather than in the associated ribosomal proteins. It is also involved in binding elongation factor G, which is required for translocation.

6.3.2 Function of the Ribosome Cycle in Translation

Polypeptide synthesis can be divided into three stages-initiation, elongation, and termination—in a process involving binding of a ribosome to messenger RNA, movement of the ribosome relative to the mRNA

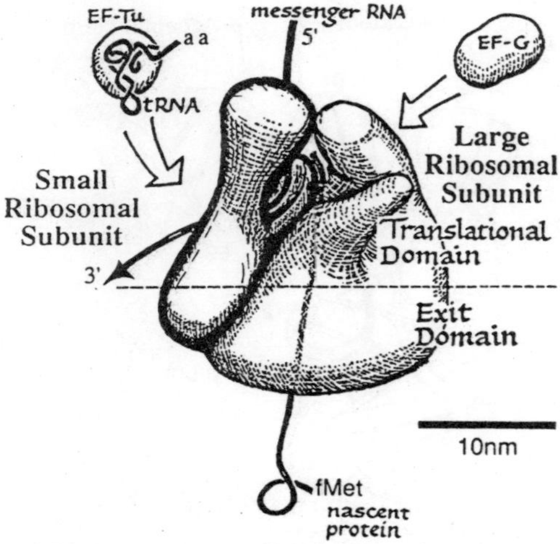

Figure 6.6: Model of the *E. coli* ribosome, based on electron microscopy. Diagram shows the relative orientation of the large and small subunits and other functional components involved in polypeptide chain synthesis.

in the 5′ → 3′ direction, and release of both mRNA and the finished polypeptide from the ribosome.

The ribosome cycle starts with the stepwise formation of an initiation complex. First, charged initiator tRNA (tMet-tRNA, or Met-tRNA$_i$) binds to the small ribosomal subunit in the P site. Next, mRNA is bound to the small subunit by an interaction involving base pairing between the initiation codon of the mRNA and the anticodon of the initiator tRNA.

Last, the large ribosomal subunit is joined to this preinitiation complex. After entry of the appropriate aminoacyl-tRNA (as selected by base pairing with the second codon of the mRNA) into the A site of the initiation complex, the first peptide bond is formed by transfer of the tMet residue from the initiator tRNA to the amino group of the aminoacyl-tRNA in the A site.

This reaction is catalyzed by the peptidyl transferase activity of the large ribosomal subunit. After formation of the peptide bond, the newly synthesized peptidyl-tRNA is translocated from the A to the P site, together with movement of the mRNA relative to the ribosome through a distance equivalent to one codon. At the same time, the free tRNA in the P site is moved into the exit or E site before leaving the ribosome.

Repetition of this process of elongation and translocation results in stepwise growth of the polypeptide chain by one amino acid residue at a time. Simultaneously, the ribosome moves one codon at a time toward the 3′ end of the mRNA until a termination codon (UAA, UAG, or UGA) enters the A site, whereupon a release factor is bound and catalyzes the hydrolysis of the peptidyl-tRNA bond.

The finished polypeptide chain is released, and the ribosome dissociates into subunits with the release of both the final tRNA and mRNA. In the ribosome cycle the same mRNA molecule and the other components needed for protein synthesis can be used repeatedly. The various steps of the translation process are facilitated by protein factors, and usually several ribosomes are attached to one mRNA molecule at the same time, giving rise to polysomes (polyribosomes).

6.4 CONCLUDING REMARKS

The polypeptide chains of all proteins are synthesized by the process just described. This mechanism gives rise to primary polypeptide chains, which are often further modified for example, by cleavage into smaller peptides, by structural modification of selected amino acid residues, or

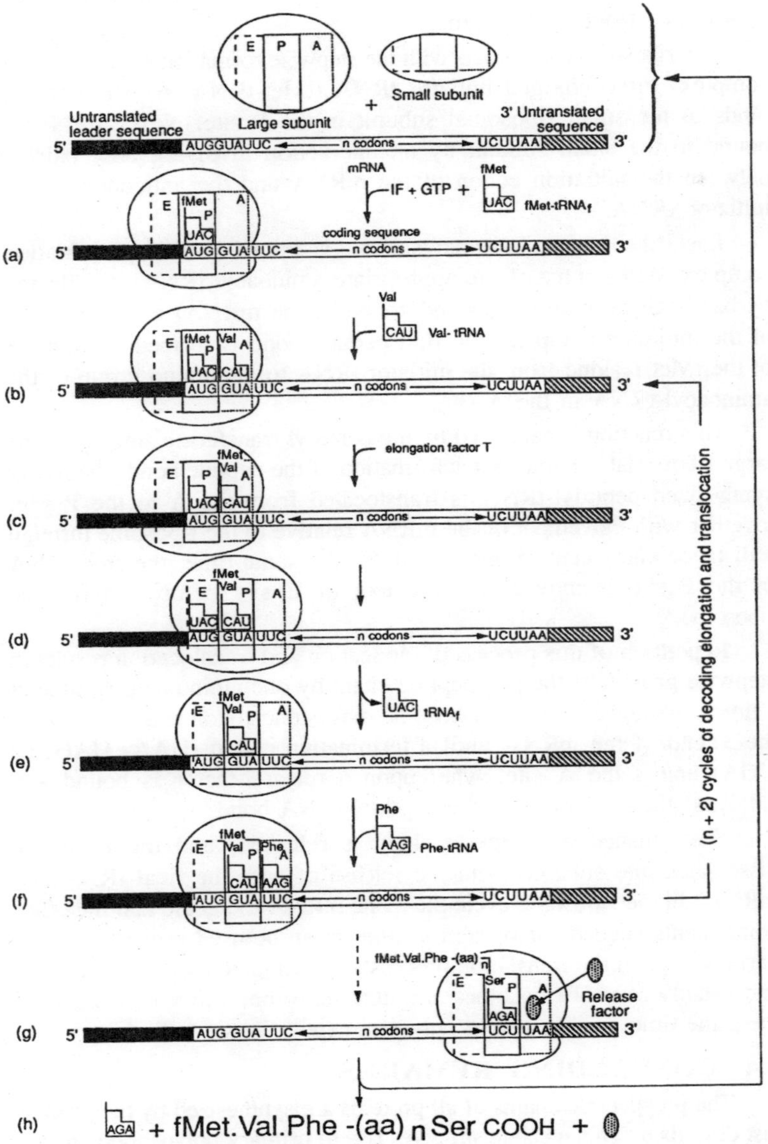

Figure 6.7: Schematic view of the biosynthesis of a polypeptide by translation of prokaryotic mRNA.

by the formation of covalent bonds between polypeptide chains. Some of these secondary modifications are related to the folding of polypeptides to the correct secondary and tertiary structure.

The production of active enzymes or peptide hormones from inactive precursors (e.g., insulin from proinsulin) also involves modifications of the primary polypeptide. Furthermore, the transport of proteins within the cell or the secretion of extracellular proteins is often linked to structural changes in polypeptide chains either during synthesis or after its completion.

The control of protein synthesis, either by regulation of the amount of mRNA available for translation or by the efficiency with which it is translated, is important in cell growth and development as a factor determining the level of cellular and extracellular proteins.

Subversion of this control occurs in cells infected by viruses: the viral nucleic acid uses the protein-synthesizing machinery of the host cell, thereby changing normal cell metabolism in favour of the synthesis of viral proteins needed for the production of virus progeny.

7

Laboratory Methods

An ongoing international effort to map and sequence the human genome will have a profound effect on future biological research and medical practice. Genes responsible for inherited diseases will be discovered and diagnostic procedures implemented.

A better understanding of gene organization and regulation will add to our understanding of human development and aging. However, sequencing the 3 billion nucleotide bases that make up the human genome is a monumental task that will require a significant expenditure of time and money.

Therefore the development and evaluation of labor-saving and cost-effective new technologies for sequencing DNA, such as scanning-tunneling microscopy, is an important issue.

7.1 STM OPERATION

Scanning-tunneling microscopes (STMs) measure the topographical profile of samples and, unlike conventional light or electron microscopes, do not require transmitted waves or lenses for their operation. Imaging is accomplished by detecting changes in the electronic tunneling current between a sharpened tip (mounted on a piezoelectric crystal) and a conducting sample as the tip is serially scanned over the sample surface.

A measurable tunneling current is generated when the wave functions of the electrons on the tip have sufficient overlap with the wave functions of the electrons on the sample surface. By applying a difference in electrical potential between the tip and the sample, a tunneling current can be maintained. There is a strong exponential dependence of the tunneling current on the distance between the tip and the sample.

When a piezoelectric crystal is used to position the tip within a nanometer of the surface, the current changes by an order of magnitude if the gap between the tip and sample changes by only 0.1 nm. The effect of this strong gap-dependent current is to concentrate the current into a very tiny region only 0.1 to 0.5 nm wide.

For electrical conductors and semiconductors, which have abundant charge carriers, atomic resolution is routinely obtained if the surface is flat. Deviations from flatness on a nanometer scale cause switching of the tunneling current to other regions of the tip during a scan.

As a result, the tip profile smears local topographical data and degrades the resolution. These effects on the signal set limitations on the topographical relief of samples, a fact that must be recognized in sample selection and image treatment.

7.1.1 Conductivity

Electrical conductivity of the sample is another limitation that must be considered in the selection of biological samples for scanningtunneling microscopy. Inevitably, biomolecules tend to be either electrical insulators or poor semiconductors, and this property poses a problem when charged particles are used as probes.

If the molecule cannot bleed away excess charge sufficiently, it will distort the results of probing by electrons. Furthermore, electrical insulators typically do not dissipate heat readily, and energy deposited by any probe can cause disintegration.

This problem is overcome in conventional scanning electron microscopes by coating samples with a thin metal film; since, however, the highest possible resolution will be necessary to sequence DNA in the electron STM, uncoated samples must be placed directly on a conducting substrate.

Electron tunneling can occur through an insulator as thick as 4 to 5 nm, and since the DNA double helix is only about 2 nm thick, a tunneling image can be obtained. However, if the reduction in tunneling current is too great, the electronic feedback system will plunge the tip into the sample. Special precautions must be taken to understand what is happening in a given case.

7.1.2 DNA Imaging

Because deoxyribonucleic acid molecules contain the genetic code for most living organisms, including man, considerable interest has arisen in the scanning-tunneling microscopy of DNA. One primary objective is to determine the nucleotide base sequence and therefore

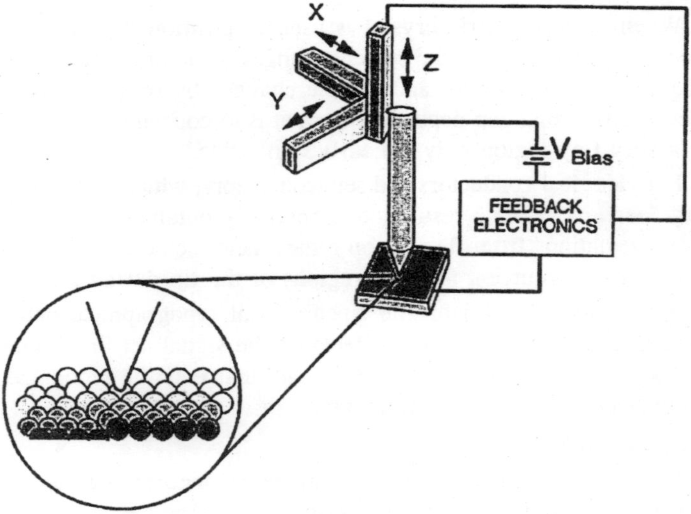

Figure 7.1: Schematic of a scanning=ntunneling microscope. While a constant tunneling current is maintained, an atomically sharp tip is scanned over a sample surface and an electronic image is created.

rapidly and accurately read the genetic code. As a starting point for imaging, the DNA molecules must be placed on a flat conductive surface. Historically, atomically flat highly ordered pyrolytic graphite (HOPG) or gold surfaces were used.

However, these surfaces are crystalline and usually chemically inert. Therefore, they lack sites to hold DNA strongly enough to resist removal by the physical and electronic forces exerted by the tunneling tip. A secondary problem initially encountered in imaging DNA on these surfaces was that structural defects intrinsic to the substrates could be confused with DNA, since both were seen only occasionally.

This was especially true when short fragments of DNA were being imaged, and on surfaces such as HOPG, which bear some irregularities that closely mimic DNA. However, DNA molecules can be positively identified by exploiting differences in conductivity between DNA and mounting surfaces.

A dc bias voltage between the tip and the sample is typically applied to maintain tunneling and acquire a topographic image. It is also electronically possible to superimpose an ac bias voltage between the tip and the sample and simultaneously record the derivative of the tunneling current with respect to the applied bias voltage.

This procedure yields an image of the local electronic conductance of the material directly beneath the probe tip. There is a large difference noted between DNA and surface artifacts.

7.1.3 STM Techniques

Both chemical and electrochemical techniques have been used to attract and hold DNA to surfaces. One of the more successful techniques, which resulted in the first STM images of entire genetically functional plasmid molecules, involves the coulostatic immobilization of DNA onto chemically modified gold surfaces.

To this end, thiol organic compounds are reacted with the gold substrate, and positively charged amino groups on the other end serve as molecular tethers to attract and hold the negatively charged DNA molecules.

Such techniques immobilize the biomolecules so well that the DNA is observed at the surface coverage expected, and confusion with rare artifacts is eliminated.

7.2 STM PROGRESS ACHIEVED

7.2.1 Direct Sequencing

Although significant progress has been made in imaging DNA with the scanning-tunneling microscope, sequencing by STM has not been accomplished. Direct sequencing would require denaturation of double-stranded DNA and the imaging at very high resolution of individual bases on one of the resulting single strands.

Isolated bases have been imaged by STM with sufficient molecular resolution to identify the double-ring structure of the purine bases (adenine, guanine) and the single-ring pyrimidine bases (thymine, cytosine). Though these results are encouraging, it would require atomic resolution to differentiate topographically between the purine and the pyrimidine bases.

For example, separating cytosine from thymine would depend on recognizing the subtle differences of ring positions 5 and 6, since thymine has a methyl group at position 5 and a carbonyl oxygen at position 6, while cytosine has an amino group at position 6. By either physically or chemically tagging one purine and one pyrimidine base, however, all four bases could be identified topographically at molecular resolution.

7.2.2 Secondary Techniques

Alternatively, it may be possible to employ secondary techniques along with topographic imaging to identify nucleotide base sequences. We have already described how topographic and differential conductivity images can be generated simultaneously to differentiate between poorly conductive DNA molecules and surface artifacts.

This technique could be refined and extended to the level at which differences in topographic images of the bases could be combined with additional high resolution electronic spectroscopy to perhaps detect conductivity differences in the bases so that the sequence could be visualized through contrast changes.

The incorporation of secondary techniques into scanning probe microscopes, for sequencing DNA, may be extended to the development of hybrid instruments. The photon scanning-tunneling microscope (PSTM), an instrument developed in our laboratory, is analogous to the scanning-tunneling microscope except that electrons are replaced by photons.

Instead of detecting the electron tunneling current between an atomically sharp conducting tip and a sample surface, as the tip is scanned over the surface, the PSTM uses a transparent probe tip to collect the frustrated total internal reflection of photons in a sample.

The photons are converted to an electrical signal by a photomultiplier tube and processed by the STM control electronic package. Although the spatial resolution of the PSTM is roughly tens of nanometers for visible light, and not as good as the STM, the optical spectroscopic resolution of the PSTM is better than electron spectroscopy by a factor of nearly 10^6.

This instrument demonstrates that the wavelength limit of conventional microscopy is not a fundamental limitation. Therefore, a hybrid instrument combining the high resolution imagery capabilities of STM with sensitive optical spectroscopy of fluorescence to a local optical probe may be possible. This challenging new area of research, combining other analytical techniques to probe microscopies, has attracted the attention of many laboratories.

The near-field optical microscope has been improved to follow surface topography and record optical data at a resolution of better than 20 nm. The combination of atomic force microscopy with optical capabilities has already been described, and instruments that can be used to map either fluorescently labeled homologous nucleotide sequences or site-specific proteins to DNA, or even to sequence DNA rapidly and accurately may be developed and perfected.

7.3 PCR TECHNOLOGY

The polymerase chain reaction (PCR) is an in vitro method for the primer-directed enzymatic synthesis of millions of copies of a specific

DNA segment. The reaction is based on the annealing and the extension by a thermostable DNA polymerase of two oligonucleotide primers that flank the target DNA segment.

The ability to amplify a specific DNA sequence from a complex template such as human genomic DNA in a I-2 hour automated reaction has allowed PCR to supplement or even replace many of the traditional molecular cloning applications in basic molecular biology research and has introduced DNA analysis into clinical diagnostics.

The PCR-based analysis of allelic sequence variation has been applied in HLA tissue typing for transplantation, in genetic typing for individualization of forensic evidence specimens, and in the diagnosis of genetic disorders such as sickle cell anemia, cystic fibrosis, and the fragile X syndrome.

PCR has also been applied to the detection of infectious disease pathogens, such as the human immunodefieieney virus and Chlamydia, and, in cancer diagnostics, to the detection of inherited as well as somatic mutations in oneogenes and tumor suppressor genes.

The ligation onto any collection of DNA fragments of short DNA sequences that can serve as primer-binding sites confers on that collection the ability to he replicated by PCR, allowing for selective procedures and *in vitro* evolution.

7.3.1 Principles

The polymerase chain reaction (PCR) is an *in vitro* reaction developed by scientists at the former Cetus Corporation in the mid1980s for the enzymatic synthesis of millions of copies of a specific DNA segment from a complex template.

The reaction is based on the annealing and extension of two oligonucleotide primers that flank the target region in duplex DNA; after denaturation of the DNA, achieved by heating the reaction, each primer hybridizes to one of the two separated strands such that extension from each 3'hydroxy end is directed toward the other.

The annealed primers are then extended on the template strand by using a DNA polymerase. These three steps (denaturation, primer binding, and DNA synthesis by primer extension) constitute a single PCR cycle. If the newly synthesized strand extends to or beyond the region complementary to the other primer, it can serve as a primer-binding site and template for subsequent primer extension reactions.

Consequently, repeated cycles of denaturation, primer annealing, and primer extension result in the exponential accumulation of a discrete fragment whose termini are defined by the 5' ends of the primers.

This exponential amplification occurs because, under appropriate conditions, the primer extension products synthesized during a given cycle function as templates for the other primer in subsequent cycles. Thus, each cycle of PCR essentially doubles the amount of DNA in the region of interest.

The length of the products that accumulate during the PCR is equal to the sum of the lengths of the two primers plus the distance in the target DNA between the primers. The principle of PCR amplification is illustrated in Figure elsewhere in this chapter. PCR can amplify double-stranded (ds) or single-stranded (ss) DNA and, with the reverse transcription of RNA into a eDNA copy, RNA can also serve as a target.

Because the primers become incorporated into the PCR product and some base pair mismatches (away from the 3' end) between the primer and the original genomic template can be tolerated, new sequence information (e.g., specific mutations, restriction sites, regulatory elements) and labels can be introduced via the primers into the amplified DNA fragment.

7.4 TECHNICAL DEVELOPMENTS

7.4.1 Riagents and Protocols

7.4.1.1 Primers

The design of oligonucleotide primers to amplify a specific target DNA segment requires some knowledge of the flanking sequence information. The design of the primers should try to minimize the potential for PCR primer artifacts (e.g., the synthesis of template independent products consisting of the primers and their complementary sequences-sometimes known as "primer-dimer") by avoiding 3' sequences that are complementary to each other. Secondary structure within the primer and repetitive sequences should also be avoided.

Whenever possible, the melting temperatures of the two primers should be very similar, to ensure that a given thermal profile is optimally efficient and specific for both primers. A variety of computer programs are available to aid in PCR primer design.

In some cases, the requirement for flanking sequence information can be overcome by ligating a fragment of known sequence to template DNA fragments to serve as a priming site. NCR with random sequence primers can also be useful in amplifying DNA for some purposes.

7.4.1.2 Polymerases

The initial studies that relied on the PCR to amplify specific targets (e.g., (3-globin) from human genomic DNA utilized the Klenow fragment of *Escherichia coli* DNA polymerase 1.

Although widely used by molecular biologists, this polymerase is not a thermostable enzyme, and its inactivation at the high temperatures necessary for strand separation required the addition of enzyme after the denaturation step of each cycle.

This requirement was eliminated by the introduction of a thermostable DNA polymerase, the *Taq* DNA polymerase, isolated from the thermophilic bacterium *Thermus aquaticu.s*. The use of *Taq* DNA polymerase transformed the PCR by allowing the development of simple automated thermal cycling devices for carrying out the amplification reaction in a single tube containing all the necessary reagents.

The availability of thermostable polymerases has not only simplified the procedure for the PCR but has increased the specificity and yield of the amplification reaction. The incorporation of *Taq* DNA polymerase into the PCR protocol allowed the primers to be annealed and extended at a temperature much higher (e.g., 60°C) than was possible with the Klenow fragment (e.g., 37°C), eliminating most of the nonspecific amplification.

Moreover, longer PCR products could be amplified from genomic DNA, probably because there was a reduction in the secondary structure of the template strands at the elevated temperature used for primer extension with a thermostable polymerase. Since the introduction of *Tay* polymerase, a variety of thermostable DNA polymerases from thermophilic or hyperthermophilic bacteria have been isolated, characterized. and used in the PCR.

Some of these polymerases, unlike the *Taq* polymerise, have 3'- to 5'exonuclease activity (the so-called editing or proofreading function). The cloning of the genes encoding some of these enzymes has permitted the production of genetically engineered variants of these native polymerises, such as a mutant of *Taq* polymerase that lacks the 5'- to 3'-exonuclease activity.

In general, polymerases that have or lack the 5' to -3'- and the 3'- to 5'-exonuclease activities will perform differently in the PCR. For example, polymcrascs with a 3'-to 5'-exonuclease activity probably will not perform well in the PCR strategy known as allele-specific amplification of sequence-specific priming based on the reduced efficiency of extension for a primer that is mismatched with the template strand at the 3' end.

This limitation of such polymerases is due to the removal of the

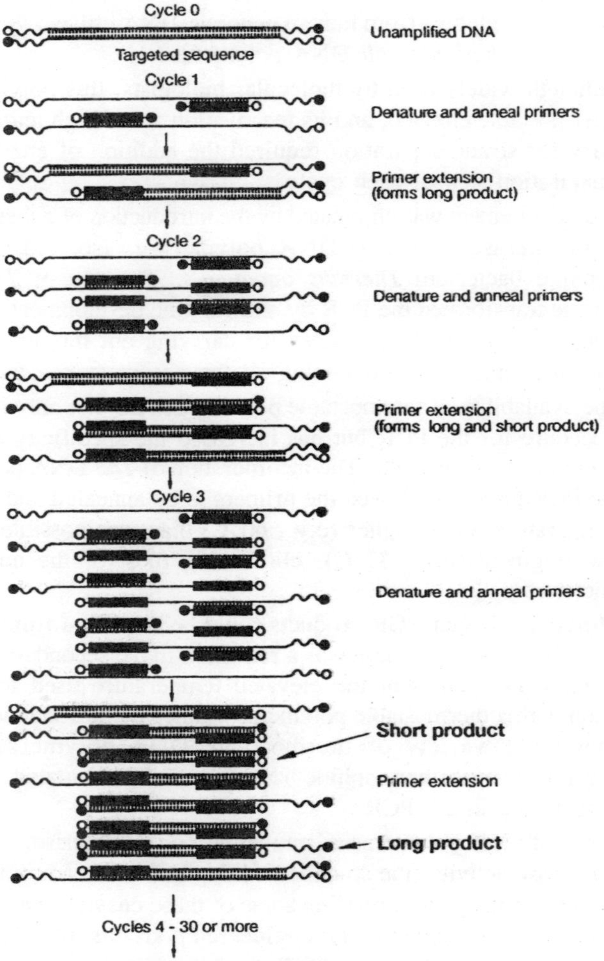

Figure 7.2: Polymerase chain reaction amplification cycles: open circles. 5' or phosphate end; solid circles. 3' or hydroxyl end of polynucleotide sequence.

mismatched base at the 3' end of the primers, creating a matched 3' end that can be efficiently extended. On the other hand, these polymerases have a greater fidelity in DNA synthesis as a result of their ability to remove misincorporated and thus mismatched nucleotides.

Many of the new thermostable polymerases have additional useful activities. The thermostable DNA polymerase from *Thermos thermophil/ us (7th)* can reverse transcribe RNA efficiently in the presence of

MnCl, at high temperatures. Under appropriate conditions, the DNA polymerase activity can also occur in the presence of MnCI,, allowing both cDNA synthesis and PCR amplification to be carried out in a single-enzyme, single-tube reaction.

7.4.1.3 Protocols

The initial PCR amplifications with the Klenow fragment were not highly specific; although a unique DNA fragment could be amplified approximately 200,000-fold from genomic DNA, only about 1% of the PCR product was, in fact, the targeted sequence. A specific hybridization probe was therefore required to analyze the amplified DNA.

Amplification with the *Tay* DNA polymerase, however, greatly increased the specificity of the reaction, so that for many amplifications the PCR products could be detected as a single ethidium bromide stained band on an electrophoretic gel.

Conditions that increase the stringency of primer hybridization. such as higher annealing temperatures and lower magnesium chloride concentrations, enhance specific amplification. In addition, the concentrations of enzyme and primers as well as the annealing time, the extension time, and the number of cycles, can affect the specificity of the PCR.

The concentration of a specific sequence in a sample can also influence the relative homogeneity of the PCR products. Thus, a single-copy nuclear gene present twice in every diploid cell usually can be detected as a unique band after gel electrophoresis of the PCR products: amplification of a sequence present in only one of 10,000 cells, however, may yield a more heterogeneous gel profile if special procedures to increase the specificity of PCR amplification are not used.

Some recently developed approaches to improve PCR specificity are based on the recognition that the *Taq* DNA polymerase retains considerable enzymatic activity at temperatures well below the optimum for DNA synthesis (72°C). Thus, during sample setup and in the initial heating step of the reaction, primers that have annealed nonspecifically to a partially single-stranded template region can be extended and stabilized before the reaction reaches the targeted 72°C temperature for extension of specifically annealed primers.

Some of these nonspecifically annealed and extended primers may be oriented with their 3'-hydroxy termini directed toward each other, resulting in the exponential amplification of a nontarget fragment. If the DNA polymerase is activated only after the reaction has reached high ($>$ 70°C) temperatures, nontarget amplification can be minimized.

This can be accomplished by an approach termed "hot start": the manual addition to the reaction tube of an essential reagent (e.g., DNA polymerase, magnesium chloride, primers, etc.) at elevated temperatures.

Thermolabile physical barriers such as wax beads that separate essential reaction components until the wax melts at high temperatures also increase specific amplification. Thus, the amplification reaction cannot proceed until the temperature in the tube is high enough to melt the wax and allow mixing of the essential reagents.

"Hot start" has been shown not only to improve specificity but also to minimize the formation of the "primer-dimcr," a doublestranded PCR product consisting of the two primers and their complementary sequences.

Once formed, this PCR artifact is efficiently amplified; it is often detected in reactions with very few or no specific templates. In rare template reactions, these short amplification products may prevent efficient target amplification by competing with the target fragment for primers and enzyme.

Another approach that can improve PCR specificity is to follow the initial amplification reaction with an additional PCR with internal, single, or double-nested primers. Like the use of oligonucleotide hybridization probes, this approach utilizes sequence information internal to the two "outer" primers to identify the subset of amplification products that corresponds to the target fragment.

The conventional nesting strategy, however, requires opening the reaction tube, eliminating or decreasing the concentration of the original "outer" primers, and then adding the "inner" primers.

Although this complicated sequence can sometimes be accomplished by simply diluting the first reaction into a second one, the procedure is both inconvenient and risky, in that it provides an opportunity for contaminating the secondary reaction with amplified product from earlier reactions.

The problem can he overcome if the outer and inner primers are all present in the initial reaction mix and if the thermal profile is programmed first to allow the outer primers but not the inner primers to amplify the targeted subset of the initial PCR products, and then to allow amplification by the inner but not the outer primers.

7.4.2 Potential Problems

7.4.2.1 Contamination

Given the capacity of PCR to generate trillions of DNA copies from a template sequence, contamination of the amplification reaction for a given sample either with products of an earlier PCR reaction (product carryover) or with exogenous DNA or cellular material can create problems both in research and in diagnostic applications.

Valuable precautions to minimize the risk of contamination include attention to careful laboratory procedures, the prealiquoting of reagents, the physical separation of the reaction preparation from the area of reaction product analysis, and the use of dedicated pipets, positive-displacement pipets, or tips with aerosol barriers that prevent contamination of the pipet barrel.

Multiple negative controls (no template DNA added to the reaction) are necessary to monitor the process and to reveal any contamination. In genetic typing, as opposed to infectious disease diagnostic testing for the presence or absence of a pathogen sequence, the contamination of a sample reaction often can he detected by a genotyping result with more than two alleles.

Recently, several approaches to minimizing the potential for PCR product carryover have been developed, all based on preventing the amplification products from serving as a template. One such strategy utilizes the principles of the restriction-modification and excision repair systems of bacteria to pretreat amplification reactions and selectively destroy DNA synthesized in an earlier PCR.

To distinguish PCR products from sample template DNA, deoxyuridine triphosphate (dUTP) is substituted fordeoxythymidine triphosphate (dTTP) in the PCR reagents and is thus incorporated into the amplification products. The presence of this unconventional nucleotide allows products of earlier PCR amplifications to be distinguished from the native DNA of the sample.

The enzyme uracil N-glycosylase (UNG), added to the reaction prior to thermal cycling, catalyzes the excision of uracil from any potential singleor double-stranded PCR carryover DNA present in the reaction prior to the first PCR cycle. The resulting abasic polynucleotides are susceptible to hydrolysis in slightly alkaline solutions (like PCR buffers) and elevated temperatures. Since UNG itself is inactivated at temperatures used in PCR, the amplification products generated during the thermal cycling are not destroyed and can accumulate.

7.4.2.2 Misincorporation and "Recombinant PCR"

During the course of PCR amplification, a mismatched nucleotide can become incorporated in the newly synthesized strand. The misincorporation rate is dependent on the reaction conditions (e.g., nucleotide concentration, pH, divalent cation concentration) and the thermostable polymerase used in the PCR. For *Tay* polymerise, the misincorporation rate is about 10^{-5} nucleotide per cycle.

Another category of very rare PCR artifacts revealed initially by sequencing cloned PCR amplification products consists of the hybrid sequences generated by in vitro recombination or template strand switching.

In amplification reactions from heterozygous individuals, hybrid sequences can, under certain conditions (e.g., very high concentrations of PCR products), result from a primer that was partially extended on one allelic template but switches to the other allelic template in a subsequent cycle.

Although misincorporation is rare and the appearance of hybrid sequences even rarer, it is nonetheless advisable to distinguish the sequence of the true genomic template from any potential PCR artifact by sequencing multiple clones derived from a single amplification reaction or by cloning the PCR products from several independent amplifications.

For most PCR applications, such as direct sequencing or oligonucleotide probe typing, it is the population of amplification products that is analyzed, and therefore rare misincorporated nucleotides are not detected.

A specific misincorporation will be present in a significant fraction of the PCR products only if the insertion of an incorrect base occurs during the first cycle of a reaction initiated with very few templates. The determination of individual cloned sequences derived from PCR products, however, can reveal such rare errors.

7.4.2.3 Recent Advances: Long PCR, Kinetic PCR, in situ PCR

About 9 years after the initial report of PCR amplification of a 110 by β-globin fragment using the Klenow fragment of *E. coli* DNA polymerase I, several modifications of the PCR procedure resulted in the capacity to amplify 30 kb fragments from genomic DNA.

Even longer fragments can be amplified from cloned templates of lower complexity, such as insert DNA in X, cosmid, and yeast artificial chromosome (YAC) clones; for some of these vectors, PCR is able to amplify the whole insert sequence.

These developments should have a major impact on cloning and sequencing strategies and are expected to expand even more the critical role played by PCR in molecular biology research. Changes in the DNA extraction protocols and modified buffers prevented singlestrand nicks and provided longer and more intact templates for amplifying these very long fragments. Longer extension times (>15 min) and relatively high denaturation temperatures, as well as the addition of cosolvents such as glycerol and DMSO, were critical in the amplification of these longer PCR products.

In addition, specific and efficient amplification required some form of hot start. Finally, the choice of thermostable polymerase proved to be critical as well in that the long fragments could be produced only by supplementing an efficient polymerase that lacked the 3'- to 5'-exonuclease activity ("proofreading" or "editing") such as *Tay* or *Tt/I* polymerase with a small amount of another enzyme (e.g., Vent polymerase) that has this function.

The rationale for this enzyme blend is that the 3'-exonuclease activity is capable of removing misincorporated bases that otherwise would prevent the newly synthesized strand from being extended beyond the misincorporated (and mismatched) base.

Another technical advance has been the ability to monitor in real time the accumulation of PCR products by the increase in fluorescence of ethidium bromide or other DNA-binding dyes added to the PCR. Video camera monitoring of the fluorescence in the array of reaction tubes in a thermal cycler can provide quantitative data as well as a convenient diagnostic format.

The ability to amplify templates within cells or tissue sections in slides, known as in situ PCR, also represents a significant advance with broad diagnostic as well as basic research potential.

7.4.3 Instruments

In addition to the advances in the reagents and protocols for carrying out efficient and specific amplifications, new instruments for automated thermal cycling and for analyzing the PCR products have been developed. These thermal cyclers have increased rates of heating, cooling, and heat transfer to modified reaction vessels and can accommodate more samples (e.g., the 96-well array).

Some of these show a reduced thermal gradient across the heating block, resulting in more precise thermal profiles from well to well. In addition, some new models do not require the use of mineral oil to prevent evaporation and increase the rate of thermal equilibration.

Instruments to cycle temperature for amplification reactions on slides rather than in standard plastic microfuge tubes have been developed for in situ PCR. For on-line monitoring of amplification reactions ("kinetic PCR"), instruments that detect the fluorescence accumulating in reaction vessels have also been designed using either fiber-optics or video cameras. In addition, instruments for the gel electrophoretic analysis of PCR products labeled with fluorescent primers have been introduced recently.

7.4.4 Detection Formats

Many different methods for detecting the presence of PCR products or for analyzing their sequence have been developed. Detection based on the identification of electrophoretically resolved, ethidium bromide stained DNA fragments of a specific length has been a widely used method. (Combined with restriction enzyme digestion, this gel-electrophoretic method is capable of distinguishing some sequence polymorphism.)

Recently, the analysis of fluorescently labeled PCR products has been facilitated by the availability of laser-scanning instruments. These gel-electrophoretic approaches are valuable for detecting length variation (e.g., the variable number tandem repeat loci used for individual identification and/or genetic mapping) in the PCR products.

Alternative separation methods, such as high performance liquid chromatography or capillary electrophoresis, may prove valuable in the future for analyzing length variation. Some gel-electrophoretic techniques, such as denaturing gradient gel electrophoresis or the single-strand conformational polymorphism method, which is based on differential mobility in native gels caused by the secondary structure of single-stranded PCR products, arc capable of detecting some of the sequence variation in amplified DNA.

Chemicals or enzymes that cleave DNA heteroduplexes with mismatched base pairs represent another approach to analyzing sequence variation in the PCR product. If the sequence variants to be detected are known, sequencespecific oligonucleotide (SSO) hybridization probes are a valuable and widely used method for the detecting as well as the genetic typing of PCR products.

Probe-based detection is critical if the amplification reactions are not completely specific; it also eliminates the need for gel electrophoresis. Immobilizing a panel of probes in a spatial array allows the typing of the PCR product with many probes in a single hybridization reaction; this immobilized probe format has been widely used in diagnostic tests.

If the specificity of the PCR is very high, fluorescent labeling of the PCR product followed by detection of fluorescence accumulation represents a very promising approach to developing very simple, rapid, and homogeneous diagnostic tests. Some detection methods depend on the specificity of primer extension; mismatches between the template strand and the 3' end of the primer are extended very inefficiently, under the appropriate reaction conditions.

Based on this property, genetic typing methods known as allele-specific amplification or sequence-specific priming have been developed in which a PCR product is generated only if the specific allele corresponding to a specific primer pair is present in the sample.

These approaches show great potential for the detection of specific sequence variants present in rare cells, such as somatic mutations in an oncogene present in a small proportion of cells within the sample.

7.5 APPLICATIONS

7.5.1 Research

The ability to amplify specific DNA segments from a complex template such as genomic DNA or, following reverse transcription, from RNA, has allowed PCR to replace the construction and screening of recombinant genomic and eDNA libraries for many applications.

In other cases, PCR has greatly facilitated subsequent procedures, such as sequencing, oligonucleotide probe hybridization, and restriction site analysis. Given the capacity for automation of the PCR and many of these analytic methods, the genetic characterization of a large number of samples can be carried out within hours.

In addition to simplifying many basic molecular biology procedures, thus introducing the capacity for DNA sequence analysis to a wide variety of biological disciplines, PCR has made possible several new experimental approaches to research questions based on the ability to amplify specific segments from very small samples (e.g., individual sperm).

In cell biology, gene expression in specific cell lineages has been analyzed by quantitative PCR methods applied to mRNA templates. These quantitative approaches typically use internal standards preceding the eDNA synthesis step; as noted, now both reverse transcription and PCR can be carried out by a single thermostable enzyme, *Tth,* from *Thermus thermophtlus.*

In molecular genetics, PCR has provided a wealth of markers for genetic mapping as well as for the construction of physical maps. In addition to genetic markers based on sequence polymorphism, genetic markers based on length polymorphism—in particular, the frequent and widely distributed simple sequence repeats (e.g., the CA repeats) have proved very valuable in constructing genetic maps based on recombination within extensive family pedigrees.

The ability to type PCR products amplified from individual sperm, which makes possible the detection of recombinant sperm, has introduced a new and very powerful approach to genetic mapping. The analysis of sequences amplified from sperm also provides a novel way of analyzing mutation.

The construction of physical maps and the ordering of cloned DNA segments have been facilitated by the identification of unique genomic sequences, known as sequence tagged sites (STSs). The presence in a given cloned fragment of these landmark STSs, each of which is associated with a specific primer pair, can be assayed by simply running a PCR with the appropriate primers.

The analysis of sequence variation, in addition to providing a rich source of genetic markers, has proved invaluable for the identification of specific mutants in critical disease-associated genes-for example, *CFTR* (the cystic fibrosis transmembrane conductance regulator gene)-and has provided significant insights into structure-function relationships.

This approach, which offers a relatively easy way to obtain DNA sequence information from a variety of individuals, has revolutionized molecular evolutionary studies; phylogenetic analyses of the sequence database generated by PCR has allowed the reconstruction of evolutionary histories for various species, including our own.

The analysis of mitochondrial DNA sequence variation, for example, has suggested an African origin of modern humans and a migration out of Africa about 200,000 years ago. The phylogenetic analysis of HIV sequences derived from patient samples has also proved critical in understanding the natural history of HIV infection and has provided a valuable tool in epidemiological studies and the tracing of HIV transmission.

The ligation of short DNA sequences that can serve as primerbinding sites onto a collection of DNA fragments confers on that entire collection (e.g.. genomic DNA fragments) the ability to be replicated by PCR (whole-genome PCR).

If the collection of fragments to be replicated contains a random sequence segment, successive rounds of some selective procedures (e.g., binding to a given protein) can be applied and a sequence with the selected property can be isolated. This in vitro evolution approach using PCR has great potential for generating nucleotide sequences with a variety of useful properties.

7.5.2 Clinical Diagnosis

For genetic disorders, such as the autosomal recessive diseases sickle cell anemia, E3-thalassemia, and cystic fibrosis, PCR has been used to amplify the gene segments that contain known mutations, and a variety of techniques can be used for direct mutation detection.

Such strategies may involve the coamplification of several exons in a single PCR. PCR/SSO probe methods for direct mutation detection are useful for carrier screening as well as for prenatal diagnosis for such diseases, provided the majority of mutations have been characterized.

A recently developed format in which the PCR product, labeled with biotinylated primers, is hybridizcd in a single reaction to an array of probes immobilized on a nylon membrane is used in PCR-based genetic tests for forensic proposes and for human individualization, for HLA tissue typing for transplantation, and for detecting mutations in the CFTR gene.

The detection of infectious disease pathogens using PCR amplification of a pathogen-specific DNA segment is a very sensitive and powerful diagnostic test. Many PCR-based infectious disease tests are offered as service tests from clinical laboratories, and commercial test kits are now available. PCR-based tests for the presence of HIV-I proviral DNA or RNA offer many advantages over conventional diagnostic tests, such as early detection (due to increased sensitivity).

Unlike serologic tests, moreover, PCR-based tests are valuable in assessing mother-child transmission. Quantitative PCR tests for HIV RNA, based on reverse transcription of the viral RNA into DNA, are useful not only in clinical diagnosis but in evaluating the efficacy of new drugs. Many other PCR tests for pathogenic fungi and bacteria, as well as DNA and RNA viruses, are currently performed from a variety of clinical samples (blood, urine, sputum, etc.).

There is enormous clinical potential available in the capacity of PCR to detect rare somatic mutations in oncogenes or tumor suppresser genes as well as some of the cancer-associated translocations. Early detection of cancer cells by sampling of sputum, urine, blood, or stool sample is a very exciting possibility.

The use of a strategy known as allele-specific amplification, in which the PCR primers amplify the rare somatic mutation selectively, represents an extremely sensitive approach. Already, the ability of PCR to detect specific chromosomal translocations (e.g., the *bcr-abl* gene fusion resulting from the translocation known as the Philadelphia chromosome, associated with chronic myeloid leukemia) has been applied to monitor residual disease in treated patients in many clinical reference laboratories. The identification of inherited mutations, such as in the p53 gene, the retinoblastoma locus, or in the ***recently identified HNPCC (hereditary nonpolyposis colon cancer)*** locus can also help identify individuals at high risk for cancer.

7.5.3 Forensic Science

Conventional genetic markers, like blood group serologic types or protein electrophoretic variants, cannot be applied to many forms of' forensic evidence. In addition, the use of DNA fingerprinting, the analysis of genomic length polymorphisms (variable number tandem repeats, VNTRs) based on Southern blotting, typically requires reasonably large amounts (> 100 ng) of intact, high molecular weight DNA.

PCR, however, is capable of analyzing minute quantities of degraded DNA, and this method is now widely used to analyze the genetic patterns of polymorphic DNA sequences amplified from biological evidence found at crime scenes. One PCR-based test, a reverse dot-blot test for polymorphism at the HLA-DQa (DQA/) locus (HLA-DQa Amplitype kit), has been applied by many forensic laboratories to the analysis of thousands of cases since Penisvlow,a vs. Pestinikis in 1986, the first case in which such PCR results were admitted as evidence.

For most of these evidence samples (typically > 5 evidence specimens/case), DNA typing based on RFLP analysis could not provide a genotyping result. In general, PCR typing of evidence samples for individual identification can be based either on the analysis of length polymorphism (VNTR regions or di , tri-, or tetranucleotide repeats, known as short tandem repeat, STR, regions) or sequence polymorphism.

The amplification of VNTR or STR regions can be carried out by using primers complementary to unique sequences that flank the tandem repeat regions, and allelic variants can be distinguished by determining the length of the PCR products amplified from the sample.

In addition, if the two alleles determine PCR products of very different lengths, the smaller fragment can, under conditions of limiting DNA polymerise, be preferentially amplified. In addition, the larger allelic fragment may be preferentially lost when amplifying a VNTR polymorphism from a partially degraded DNA sample.

For both VNTR and STR region markers, the definition of alleles in the sample is made using an allelic size ladder, a pool of PCR products comprising the range of allelic size variants in the population. In some cases, however, sequence polymorphism may also he present in VNTR regions, making the definition of allelic length variants by electrophoretic mobility difficult. One recent approach to PCR-based genetic typing, the *minisatellite variable repeat (MVR) approach, utilizes both the* length and sequence polymorphism in some VNTR regions for purposes of individualization.

Sequence polymorphisms can be detected by a variety of methods including the use of oligonucleotide hybridization probes. With the use of the immobilized probe/reverse dot-blot method, in which a panel of oligonucleotide probes is immobilized in a spatial array on a nylon membrane, the complexity of the test is independent of the number of probes because the amplified sample is analyzed with *all* the probes in a single hybridization reaction.

As already noted, this format was developed for typing *the polymorphism at the HLA-DQA / locus; it was the first commer*cially available PCR-based genetic typing kit. If the genotype of the evidence sample differs from that of a suspect's reference sample, that person is excluded or eliminated from the pool of potential donors of the evidence.

If the genotypes of the suspect and the evidence sample match, the possibility that the suspect is in fact the donor cannot be eliminated. The significance of this match for identifying the source of the evidence specimen is a function of the frequency of this specific genotype, as determined by population surveys.

Similar dot-blot tests for determining the sequence polymorphism in the D-loop segment of mitochondrial DNA have recently been developed and applied to the forensic analysis of bone fragments. A recently developed genetic test is based on the coamplification of five additional marker loci (each of which has two or three alleles) and HLA-DQa and analysis of the PCR products by the reverse dot-blot method.

If the DNA type of multiple loci of an evidence sample matches that of the suspect, the probability that a random individual in a given population would have the same genetic pattern can be estimated, assuming statistical independence, by multiplying the probabilities of the individual loci.

Clearly, as more PCR-based genetic markers are developed and used, the potential for individualization will increase dramatically. The same genetic tests for individual identity testing in forensic evidence specimens have also been used in paternity testing as well as in the monitoring of bone marrow engraftment by distinguishing donor and recipient cells. Potential mix-ups of cell lines and clinical specimens have also been detected quickly by simple PCR-based assays.

7.6 PERSPECTIVES

In the decade or so since the first report of PCR amplification and its use for genetic diagnosis in 1985, the number of different applications has grown dramatically.

The technology of PCR amplification has also evolved over this period as new polymerases and new protocols have increased the efficiency and specificity of PCR as well as the length of amplified DNA fragments.

With this recently developed capacity, PCR is likely to replace conventional molecular cloning strategies even more than it has in the past. In addition, new developments such as in situ PCR promise to have a significant impact in both research and diagnostic applications. In general, PCR has made the information embedded in nucleotide sequences more accessible to the basic researcher and clinician alike.

8

Lignin Genetic Engineering

Over the last 35 years, the production of paper has more than tripled. To be competitive, to reduce the uncontrolled exploitation of wild forests, and to limit energy and environmental problems, the forest products industry must develop genetically improved woody materials. Until now, genetic improve ment of trees was achieved using conventional breeding methods, which are particularly difficult to exploit for woody species because of long generation times and evaluation of specific characters at the adult stage.

Among the various facets of plant biotechnology, genetic engineering offers new opportunities and attractive perspectives for improving woody species. The first transgenic tree was obtained in 1987. Since that time, the development of genetic transformation techniques for several tree species and progress in the characterization of several important genes in woody plants have opened the door to wide exploitation of genetic engineering in forest management.

Genomic and genetic transformations are entering the field of plant biology at an extremely rapid pace, and their impact should enhance understanding of plant function and lead to the improvement of crops and forest products. This may correspond to increases in productivity but also to optimization of the chemical composition to provide new products better suited to nutritional, pharmaceutical, or industrial purposes. As the composition of wood is very important for the pulp industry, lignin genetic engineering is a very active area of research that has been stimulated in recent years by the characterization of important genes controlling lignification.

A significant number of transformed plants exhibiting qualitative and quantitative changes in their lignins have already been obtained and, in a limited number of cases, preliminary data have demonstrated the industrial usefulness of the resulting products.

Pulp and paper production requires the use of costly, energy-consuming, and often polluting treatments to separate lignins from cellulose, and it is clear that genetically engineered trees that produce modified lignins (in lower amounts or that can be more easily processed) would lead to substantial industrial and environmental benefits.

The successful extension to clonal forestry of the efficient strategies defined on model systems has at least four prerequisites: convenient germ plasm, large-scale vegetative propagation, convenient transformation techniques, and long-term in-field confirmation of the effects observed in model plants.

Investigations are being actively conducted along these different lines and, in parallel, investigations are conducted on easy-to-transform model plants in order to define the best ways to engineer new cell wall compositions. This chapter will critically highlight recent results in this area, suggest new strategies for the future, and emphasize the long-term benefits that can be expected from manipulating lignin metabolic pathways.

8.1 RECENT ACHIEVEMENT IN THE FIELD OF LIGNIN GENETIC ENGINEERING

Among the polymers that contribute to the plant cell wall, little has been done with genetic manipulation of cellulose and hemicelluloses because of a lack of convenient regulatory genes. In contrast, over the last decade, it has become clear that genetic engineering can tackle lignin composition and content.

This is due to a reasonable knowledge of the biosynthetic pathways (and corresponding enzymes) leading to lignins even though recent results have revealed unexpected new pathways or new regulatory circuits. Figure elsewhere in this chapter summarizes our present knowledge of lignin biosynthetic pathways.

It is clear that the synthesis of lignins from phenylalanine proceeds through three different sets of reactions.

1. The common phenylpropanoid pathway, which provides general intermediates (cinnamoyl-CoAs) for different phenolic syntheses
2. The monolignol specific pathway, which gives rise to the three building units of the lignin polymer, the coumaryl,

coniferyl, and sinapyl alcohols

3. The polymerization of monolignols, for which peroxidases and laccases are candidates

Most of the genes encoding the enzymes along these pathways have now been characterized with the exception of the hydroxylase converting coumaroyl-coenzyme A (CoA) into caffeoyl-CoA. These genes have been exploited through antisense or cosuppression strategies in order to reduce the activity of the corresponding enzymes and alter the functioning of the lignin pathway. The modifications of lignin profiles that have been envisaged until now have dealt either with reduction in lignin content or with changes in the lignin monomeric composition because this parameter is important for lignin extractability.

However, it soon appeared that experiments aiming to decrease the lignin content indirectly altered the chemical composition of transgenic lignins. In the following sections significant results obtained through the down-regulation of different target genes will be highlighted irrespective of the initial objective.

8.1.1 Down-regulation of O-Methyltransferases

Higher plants contain at least three classes of O-methyltransferases that can control the degree of methylation of the monolignols. Genetic engineering experiments have confirmed that caffeic acid 3-O-methyltransferases (COMTs) give rise to sinapic acid from 5-hydroxyferulic acid and that caffeoyl-CoA 3-O-methyltransferases (CCoAOMTs) lead to feruloyl-CoA.

In addition, other methyltransferases can probably convert hydroxyconiferaldehyde and hydroxyconiferyl alcohol into sinapaldehyde and sinapyl alcohol, respectively. The down-regulation of COMT both in tobacco and in poplar through the antisense strategy did not decrease the lignin content, whereas the monomeric composition was dramatically modified. Transgenic lignin has a decreased S content while the amounts of G units remained relatively unchanged.

In addition, these transformed lines accumulated a novel 5-hydroxyguaiacyl unit in their lignins. Chemical and pulping experiments revealed that these lignins contain, as could be predicted, more condensed bonds (C-C) and are more difficult to extract from the cell wall (higher kappa number) than the control lignins.

These transformed plants are not particularly of interest to the pulp industry but may be of interest if the wood is used for fuel because the calorific value of condensed lignins should be improved. Several groups have envisaged down-regulating another methyltransferase (CCoAOMT)

considered to be particularly involved in the synthesis of coniferyl alcohol. CCoAOMT has been shown to be encoded by multigene families in different plants including poplar and tobacco, and specific members of these multigene families have been used in antisense or cosuppression experiments.

Reduction of CCoAOMT alone in transgenic tobacco plants resulted in a decreased lignin content but a simultaneous reduction in CCoAOMT and COW activities induced a further reduction in lignin content, confirming that both enzymes are indeed involved in methylation reactions in lignin biosynthesis.

Surprisingly, although the transgenic plants showed a 40 to 60% reduction in lignin content, they appeared to grow normally under greenhouse conditions even though they exhibited a deformation of vessel elements. These results have been partly confirmed by M. Legrand's group in Strasbourg (France) (personal communication) on the same material.

However, in addition to a reduction in lignin content, CCoAOMT downregulation induced a decreased growth rate and a dramatic disorganization of vascular tissues (reduction of xylem thickness, reduction of vessel diameter). These discrepancies concerning the growth of the plants have not yet been explained. W. Boerjan et al. (personal communication) obtained poplar lines cosuppressed for CCoAOMT that exhibit a slight reduction in lignin content.

It thus appears that CCoAOMT alone or in combination with COMT is an interesting target for lignin genetic engineering, particularly if the developmental effects observed for some transgenic lines can be reduced. From a qualitative point of view, G units were preferentially reduced in the β-O-4-linked monomer fraction of these transgenic plants, leading to an interesting increase in the S/G ratio of the lignins.

8.1.2 Down-regulation of 4-Coumarate CoA Ligase (4CL)

This enzyme governs a key step of the common phenylpropanoid pathway leading to hydroxycinnamoyl CoAs. A down-regulation of its activity should lead to pleiotropic effects, such as those observed when phenylalanine ammonia-lyase (PAL) and cinnamate-4-hydroxylase (C4H) are down-regulated, if the targeted enzymes (isoforms) are not strictly dedicated to monolignol biosynthesis.

However, interesting results have been obtained in *Arabidopsis,* tobacco, and poplar 4CL down-regulated plants, suggesting that this enzymatic step is an interesting candidate for lignin genetic engineering. In both *Arabidopsis* and tobacco, a nearly total block in 4CL activity led to only a modest reduction in total lignin content and the plants were morphologically normal. In contrast, transgenic aspen exhibited a substantial reduction in lignin quantity and a 15% relative increase in cellulose content.

In addition, tree growth was substantially enhanced and phenolic profiles of the upper leaves and shoot apex were significantly altered in transgenic lines. Even though their rationale is not understood at the moment, these secondary effects on plant growth and cellulose content are particularly interesting in the context of the pulp industry and these investigations merit further attention.

8.1.3 Cinnamoyl CoA Reductase (CCR) Down-regulation

As the first committed enzyme in monolignol biosynthesis, CCR channels phenylpropanoid metabolites into the biosynthesis of lignins. Significant down-regulation of CCR activity in tobacco was obtained by ectopic expression of the homologous tobacco antisense gene.

The CCR downregulated tobacco plants with a moderate decrease of CCR activity exhibited only a very slight reduction in lignin content and had a normal phenotype, whereas the most severely inhibited transformant showed a 50% reduction in lignin content and an abnormal, heritable phenotype (reduced growth, abnormal leaf morphology).

In these plants, the hypolignified xylem vessels were unable to withstand the compressive forces generated during transpiration and tended to collapse inward. In these transformants, the yield of thioacidolysis products was reduced, reflecting a lower proportion of β-*O*-4 bonds in lignin.

However, this uncondensed lignin fraction was enriched in S units. A substantial increase in G units containing free phenolic groups was also observed (C. Lapierre, personal communication). In addition, an increase of cell wall-linked phenolics released by mild alkaline hydrolysis occurred, the main enrichment concerning ferulic and sinapic acids and acetosyringone.

It was suggested that the incorporation of ferulic acid into the cell wall was responsible for the brown-orange colouration observed as a consequence of CCR silencing. Additional studies using ^{13}C nuclear magnetic resonance (NMR) revealed the presence of ferulate tyramine in cell walls of the CCR antisense line.

These different phenolic compounds could be an integral part of the lignin polymer because, at least in grasses, Jacquet et al. and Ralph et al. have shown that ferulate esters can be converted to phenoxy radicals that copolymerize with lignin polymers. It is clear that CCR silencing induces a marked reorientation of phenolic metabolism, resulting in a new partitioning of phenolic precursors into lignins and other related phenolic sinks. Simulated pulping experiments performed on CCR down-regulated tobacco plants obtained in our laboratory or in collaboration with Zeneca revealed a significant decrease in kappa number with little modification of other parameters such as cellulose yield or cellulose degree of polymerization (DP).

Taken together, these data suggest that CCR controls the entry of carbon flux into the lignin pathway. However, the resulting effects-reduction in lignin, increase in cell wall-associated phenolics and in soluble phenolics, potential decrease in other monolignol-derived compounds such as lignans and/or dehydroconiferyl alcohol glucosides-may have an impact on plant development.

Whatever the molecular mechanisms involved, it is clear that these results illustrate the unexpected effects that can be associated with a dramatic decrease in lignin content (and associated responses). A compromise should thus be attained between a moderate reduction in lignin content and normal development.

8.1.4 Cinnamyl Alcohol Dehydrogenase (CAD) Down-regulation

CAD down-regulation was one of the most impressive successes of lignin genetic engineering research a few years ago. Since that time, the results initially obtained on tobacco have been confirmed by independent reports of CAD suppression in tobacco and have been extended to poplar. The plants with a strong reduction of CAD activity have normal development and a novel red colour of the xylem at certain developmental stages.

Their lignin content is not (tobacco) or slightly (poplar) altered but their lignins are enriched in coniferyl and sinapaldehydes, the aldehyde substrates of CAD. The newly formed lignins, which have been characterized in detail by NMR, are more easily extracted from antisense plant samples by sodium hydroxide or thioglycolic acid, suggesting that the structural changes in the antisense lignin make it more extractable.

Kraft pulping experiments have shown more extensive delignification of low-CAD tobacco and poplar plants when compared with control plants, demonstrating the benefit of this genetic manipulation for pulp and paper making, and field trials of these CAD down-regulated poplars are currently being performed in France and in the United Kingdom.

A confirmation of the effects of CAD down-regulation has been indirectly obtained from the study of a CAD-deficient *Pin us taeda* mutant that accumulates aldehyde components and substantial levels of dehydroconiferyl alcohol in its lignins.

Here again, the mutant plant stands normally and has a normal phenotype (except for the red colouration of xylem) and the resulting lignins are more easily extracted. These is no doubt that these CAD-deficient plants are particularly promising for the pulp industry.

8.1.5 Double Transformants

In order to combine the advantages of the down-regulation of individual genes and to yield new and useful lignins, several groups have envisaged producing double and triple transformants by crossing existing plants with single antisense genes or using transgenes containing multiple genes under the control of a single promoter.

Our recent results on CCR/CAD down-regulated plants illustrate the potential interest of this approach. By crossing homozygous tobacco lines down-regulated for CCR and for CAD, we have obtained hybrid lines downregulated for both enzymes.

If the hybrid exhibits intermediate values between the two parents for some characteristics, its lignin content is surprisingly lower than for the CCR down-regulated homozygous line. However, in contrast to this parent line, its size and morphology are not affected.

This hybrid line with only 40% of residual lignin was processed in simulated kraft pulping experiments by the Centre Technique du Papier, Grenoble (France) (M. Petit-Conil, personal communication), and a 35% decrease in kappa number, a 24% increase in yield, and a 13% increase in cellulose DP were observed.

These interesting results, which have been confirmed several times on tobacco plants obtained in culture-room conditions, should now be extended to other species grown in natural conditions. They suggest that the CCR and CAD transgenes work in synergy in a still undefined way to give rise to new lignin profiles without altering plant morphology.

8.2 POTENTIAL SIDE EFFECTS AND COMPENSATION MECHANISMS ASSOCIATED WITH LIGNIN MODIFICATIONS

One envisaged drawback related to lignin manipulation in plants could be the potential attenuation of defense mechanisms against pathogens. Indeed, lignification is considered one of the adaptive plant responses to pathogen attacks. Down-regulating lignin synthesis could reduce the ability of plants to protect themselves against pests.

Such a hypothesis has not yet been substantiated using the transgenic plants already obtained, but the use of chemical inhibitors of lignification resulted in decreasing the hypersensitive response in wheat plants challenged by *Puccinia graminis*.

In order to minimize the drop in these defense mechanisms, which usually occur at the surface of the tissues, the use of specific promoters to drive transgene expression has been envisaged. Ideally, such a promoter should be xylem specific, avoiding a decrease in stress lignins produced on the plant surfaces.

Another important aspect of lignin manipulation is the potential disturbance of plant development. Reduction or modification of the lignin content of plant cell walls may theoretically exert pleiotropic effects on plant functions through changes in the strength of plant organs, sap conduction through the xylem, or permeability of cell wall barriers (caspary band).

Changes in plant development have been observed for strongly depressed CCR tobacco lines even though it has not been clearly demonstrated whether the resulting phenotype (reduced growth) is due to changes in cell wall properties or to changes of soluble phenolic profiles. Such a reduction in size has also been demonstrated for CCoAOMT down-regulated tobacco plants.

Surprisingly, CCR/CAD double transformants exhibit a strong decrease in lignin content but no changes in morphology (at least in culture room conditions). This last observation suggests that the decrease in lignin content per se is not directly responsible for the developmental changes observed but that other unknown metabolic changes are probably involved.

It is worthwhile mentioning that in our hands the same tobacco transgenic line (CCR down-regulated) that exhibited normal growth in culture room conditions was developmentally altered when grown in field conditions. It is possible that the outside habitat provides more adverse conditions (e.g., water stress, mechanical stress) and that the transgenic line adapts itself less easily to this natural context.

It is clear that for applied purposes the transformed plants should behave as wild-type plants in terms of biomass production and agronomic performance. This has been verified for 4-year-old CAD down-regulated poplars, which appear to be promising candidates for immediate application. The same evaluation procedures at the field level should be performed in the long term for each new transformed line.

From the functional point of view, plants with altered lignin synthesis seem to adopt compensatory strategies for maintaining the integrity of their cell walls. This is clearly illustrated in the case of CAD down-regulated plants that utilize cinnamaldehydes instead of cinnamyl alcohols as building blocks of their transgenic lignins. CCR down-regulated tobacco plants partially compensate for the lack of lignin in their cell walls by a striking increase in tyramine ferulate.

Additional reinforcement of the cell wall may also be provided in these transgenic lines by a very significant increase in wall-associated phenolic compounds. The question is still open whether these new biochemical phenolic decorations of the cell wall or of the cytosolic fraction have a positive or a negative impact on the pulping process.

A negative example is provided by *Eucalyptus camaldulensis* CAD down-regulated plants that overproduce tannins, interfering with the bleaching step during the pulping process (T. Ona, personal communication). In contrast, if the compensation strategy could concern useful components of the cell wall, the benefits would be obvious.

This situation is illustrated by the exciting results obtained by Chiang's group for 4CL down-regulated aspen, for which the down-regulation of lignin, was accompanied by an increase in cellulose leading to a nearly doubled cellulose/lignin ratio. If the required defense, water transfer, and other properties survive, as they appear to in these particularly vigorous plants, the potential for improving plant utilization is enormous.

8.3 FUTURE TARGETS FOR ENGINEERING NEW LIGNINS

In addition to relatively well-defined strategies using, for example, CAD down-regulation, which can be of immediate utility, and other potential interesting target genes such as CCoAOMT, 4CL, or CCR whose interest has to be confirmed, other approaches and more diversified plant material will be exploited in the future for lignin genetic engineering. Some of the prerequisites for efficient targeted genetic modification of lignin profiles are:

- A new cell wall composition more adapted to specific agroindustrial purposes.
- A lack of secondary effects potentially inducing a reduction in plant productivity or in the efficiency of the pulping process
- Easy adaptation to plant species of high economic interest

Different strategies aiming to fulfill the first two criteria are to be envisaged in the future. Among them are the use of new target genes downstream in the lignin pathway such as laccase/peroxidase genes potentially involved in the polymerization step of lignins and/or UDPG glucosyltransferase and coniferin β-glucosidase genes potentially involved in the transport, storage, and mobilization of monolignol precursors. The exploitation of these new targets should reduce the pleiotropic effects due to a shortage in the production of monolignols.

However, it is difficult to predict whether the down-regulation of these genes would avoid, through sequential feedback mechanisms, the accumulation of intermediates upstream in the pathway. Other strategies that have been particularly successful in the case of flavonoid metabolism manipulation deal with the use of transcription factor genes regulating a whole pathway.

Genes encoding transcription factors of the MYB family are good candidates because in many of the promoters of the genes encoding enzymes of the phenylpropanoid and lignin branch pathway, binding sites for these proteic factors have been characterized. In addition, L. Tamagnone et al. have shown that ectopic expression of MYB genes in tobacco induced a reduction in lignin content.

Interestingly, A. Kawaoka and H. Ebinuma have identified a new transcription factor that controls gene expression in the lignin biosynthesis pathway (Nt lim 1). Its deduced amino acid sequence is similar to that of the LIM protein family that contains a zinc finger motif. Transgenic tobacco plants with antisense Nt lim 1 showed low expression levels of phenylalanine ammonia-lyase, 4-coumarate CoA ligase, and cinnamyl alcohol dehydrogenase and a decrease in lignin content.

This simultaneous down-regulation of several enzymes is particularly interesting because it can prevent the unwanted accumulation of phenolic intermediates along the phenylpropanoid and *lignin* pathways and facilitate the reorientation of carbon flux into primary metabolism.

The production of multiple transformants down-regulated for several genes is also interesting to fine-tune lignin profiles. This strategy is elegantly exploited by *C. Halpin* et al. through the use of multiple genes under the control of a single promoter, and in our hands it has already suggested the potential interest of CCR/CAD double transformants.

In addition to down-regulation of specific genes, the overexpression or the combined down-regulation and overexpression of key genes along the pathway may be envisaged. That is particularly illustrated by the overexpression of an angiosperm gene *F5H* in gymnosperms to increase their syringyl unit content and to make the lignin more readily extractable. Several groups are initiating research projects on this topic.

Whatever the gene or the combination of genes used, the limitation of transgene expression to specific tissues or even to specific types of cells within the lignified tissues may be an important way to avoid undesirable effects.

Until now, promoters of lignification genes such as *CCR* or *CAD* were, at least in our hands, less efficient than the 35S CAMV promoter in driving *CAD* and CCR antisense expression. However, K. Meyer et al. (30) have shown that the C4H promoter was more efficient than the 35S CAMV promoter in driving the expression of F5H in *Arabidopsis*.

These results show the potential exploitation of specific promoters for limiting the expression of the transgenes. More sophisticated approaches may be further envisaged through the use of promoters expressed in specific parts of the xylem such as fibers or vessels (*W. Boerjan*, personal communication) in order to alter fiber composition, for example, without disturbing the structure and function of the vessels.

8.4. A SOCIOECONOMIC PERSPECTIVE

World pulp production reached 198.2 million tons in 1997, and the European Union contributed 33.1 million tons to this total. It is expected that the success in lignin content reduction, in lignin solubility increase, or in lignin structure modification will have a very important impact on the pulp and paper industry.

There are a number of projects in progress around the world aiming to document the effect on trees of the modifications of lignification gene expression. One of the most advanced, partially funded by the European Union (EU), began in 1989 (OPLIGE) and gave rise to field tests of genetically engineered poplars that are under way in the United Kingdom and France.

Shell is developing an experiment with CAD down-regulated *Eucalyptus* trees in Uruguay. The trees were planted at the end of 1997 and have another year to go before they will be mature enough for felling and testing. Advanced Technologies Cambridge (ATC) in collaboration with our laboratory is also involved in the transformation of *Eucalyptus camaldulensis* with different antisense constructs containing lignification genes.

Other programs within the EU-supported project TIMBER aim to introduce such genes into spruce or into *E. globulus*, an important species for southern Europe. Additional target species for which genetic transformation appears to be realistic are larch and the *Pinus* species

Pinus radiata and *Pinus taeda,* for which efficient protocols using biolistic transformation of embryogenic tissues have been designed.

One initiative that attracted a great deal of media attention was the $60 million joint venture between Fletcher Challenge Forest, International Paper, Westvaco, and Monsanto announced in April 1999. Genesis Research and Development, Auckland, New Zealand is providing the genomic research.

The experiment will concentrate initially on eucalyptus and poplar species, radiata pine, loblolly pine, and sweet gum. The genetic improvements aimed for include herbicide tolerance, higher growth rates, and improved fiber quality in order to meet the world's wood and fiber needs without increasing pressure on native forest.

In addition to the European investments illustrated by the EU OPLIGE and TIMBER projects, this new initiative demonstrates the worldwide interest in these new biotechnologies applied to trees. In the future, it would be crucial to combine the availability of convenient genes and the ability to transform important tree species easily.

Protection of intellectual property is an important aspect of these new strategies, and several companies are launching research programs in order to characterize new interesting genes to avoid sublicensing procedures. The European groups have already filed various patents protecting the exploitation of different key genes *(COMT, CCR, CAD)* and are well advanced in the transfer of the technology to species of economic interest.

In addition to previous studies on the poplar hybrid INRA 717 B4 *Populus tremula X P alba,* easy to transform but of limited economic interest, significant results have been obtained on the economically important poplar OGY cultivar *(Populus deltoides X P nigra).*

One of the main concerns about genetically modified trees is their acceptance in public opinion. The fact that these transformed plants are nonfood products designed for industrial purposes should facilitate public acceptability. However, the recent destruction of genetically modified poplar trees at the Jealott's Hill experimental station of ZENECA near Bracknell Berkshire indicates the strong negative reactions of anti-GMO (genetically modified organisms) activists.

One of the main potential problems in addition to the unknown long-term effects is the risk of cross-pollination and the possible effect on biodiversity. In the ZENECA field trial, the trees were female and not able to produce pollen. But several approaches are being exploited worldwide for genetically engineering sterility at the same time as introducing the constructs of interest in order to avoid the release of pollen from transgenic lines.

Risk assessment for transgenic trees presents special challenges because of the space and time scale involved but also because of the complex interplay of ecological, genetic, agronomic, and social factors. Altering the lignin profiles of plants should not provide a competitive advantage to the transformed lines but could limit their resistance to different pests.

A careful examination of these new transgenic products is then absolutely necessary in order to evaluate the potential negative impacts, if any, that could outweigh the already identified advantages for the competitiveness of the pulp industry. The process of genetically modifying trees is very much in its infancy, and the public perception of the new products is a critical aspect of its wide development.

8.5 CONCLUDING REMARKS AND PROSPECTS

It is clear that the lignin content and composition of crops and woody species can be manipulated through genetic engineering. In addition to the application-orientated objectives that have triggered most of the research in this area, the impressive amount of data that have been accumulated in the last few years has generated new fundamental concepts about lignin biosynthesis pathways and lignin chemical flexibility.

As with other metabolic engineering experiments, the results obtained have also revealed pleiotropic and unexpected effects beyond the targeted pathway. Some of these could be explained through cross-talks between genes because it has been recently shown by *G. Pincon* et al. that COMT gene down-regulation has an impact on the expression of other genes of the lignification pathway.

Lignin genetic engineering is also a way to explore the subtle regulations occurring during the lignification process. The analysis of our CCR X CAD tobacco double transformants has shown that in addition to a strong reduction in lignin content, the simultaneous down-regulation of these two activities has a significant and specific impact on the nature of lignins synthesized and on the preferential cell localization of the residual lignins.

In collaboration with K. Ruel and A. Yoshinaga, it was demonstrated that the transgenic lignins were enriched in noncondensed units and that the reduction in lignin content was more important in the fibers than in the vessels. The new transformed plants may also help better identify the chemical characteristics of lignocelluloses, which are beneficial in wood processing.

Indeed, comparisons performed on wild-type and transgenic lines represent a new exploratory approach to reevaluate the criteria important for the efficiency of the pulping/bleaching process (in addition to the classical S/G ratio). Such chemical studies have been initiated by C. Lapierre (personal communication) using a range of sophisticated methods. For example, lignins from CCR and CAD down-regulated plants are enriched in free phenolic groups, which improve their extractability.

At the present stage of progress, and in addition to the evaluation of the stability of the traits in the long term, which has already been discussed, many tests still have to be performed on the existing transgenic material in order to evaluate its industrial usefulness.

Simulated pulping experiments have been performed by the CTP (Centre Technique du Papier, Grenoble, France) on tobacco stems and poplar branches and trunks of different transformants with decreased COMT, CCoAOMT, CCR, and CAD activities showing several interesting modifications in terms of kappa number, cellulose yield, and cellulose DP.

The most interesting lignin modifications should now be integrated in the pulping/bleaching process taking into account the constraints and limitations associated with each technological stage. Breeders and chemists involved in the pulp industry are usually looking for genetically engineered trees with the lowest lignin content possible but enough lignin (or the equivalent!) to enable them to withstand environmental hardships (winds, heavy rains, etc.).

At the same time, they want to process trees with lignin that would be easier to dissolve, thus requiring less chemicals, less heat, and less retention time in the digester. The resultant pulp should have equal or improved quality (attributes desired in the pulp and in the end-use product, paper) when compared with current bleached kraft market pulp. From the same biological resource, the quality of the pulp can be improved by modifying intermediate steps in the process and particularly in the cooking and the bleaching stages.

However, this improvement is usually associated with a decrease in yield and an increase in the amount of chemicals and energy required. Thus, it is clear that most of the time, a compromise has to be found between opposing objectives. An optimized resource should increase the yield and the quality of the pulp and/or decrease the quantity of energy and chemicals used in the transformation process.

Taking into account these potential limitations, the existing technology should now be rapidly extended to different woody species of economic interest in light of the progress observed in tree genetic transformation. New avenues are also open to optimize and to extend the first results.

They include, as partially described, the fine tuning of the lignin content or composition through the use of a combination of genes and the exploitation of transcription factors. Finally, progress in the characterization of the genes involved in the biosynthesis of the polysaccharides of the cell wall opens the way to targeted genetic manipulation of cellulose and hemicelluloses in plants.

A combined modification of polysaccharides and lignin profiles can also be envisaged. The resulting transformants should increase the natural variability of cell wall components in plants. They should then contribute directly or through integration in classical breeding programs to the improvement of the resource used for pulp and paper making.

9

Genetic Transformation

Plant transformation is an indispensable tool, both for the experimental investigation of gene function and for the improvement of plants either by enhancing existing traits or introducing new ones. It is now possible to introduce and express DNA stably in nearly 150 different plant species.

Many aspects of plant physiology and biochemistry that cannot be addressed easily by any other experimental means can be investigated by the analysis of gene function and regulation in transgenic plants. This offers an unprecedented opportunity to study the molecular basis of important processes that have been intractable to conventional analysis, such as the complex signal transduction pathways and hierarchies of genetic regulation that underlie plant-microbe interactions, sexual reproduction, and development.

However, much of the effort in plant transformation research reflects expectations that the technology can rapidly produce plants with improved or novel traits for the benefit of mankind. These improvements would be difficult or impossible to achieve with conventional breeding alone.

The fruits of this research are already available and include herbicide-tolerant plants; plants showing resistance to insect pests and viral, bacterial, and fungal diseases; plants with improved nutritional qualities; plants used as bioreactors to produce valuable proteins such as antibodies and vaccines; and plants in which metabolic pathways have been engineered to produce valuable products such as speciality oils and drugs.

This chapter reviews the technology available for gene transfer to plants, focusing particularly on recent advances, and discusses the constraints that must still be overcome to realize the full potential of plant transformation.

9.1 REQUIREMENTS FOR PLANT TRANSFORMATION

9.1.1 Overview

In general, plant transformation systems are based on the introduction of DNA into totipotent plant cells, followed by the regeneration of such cells into whole fertile plants. Two essential requirements for plant transformation are therefore an efficient method for introducing DNA into plant cells and the availability of cells or tissues that can easily and reproducibly regenerate whole plants.

DNA can be introduced into isolated cells or protoplasts, explanted tissues, callus, or cell suspension cultures. However, the process is characteristically inefficient and only a proportion of cells in a target population will be transformed.

These cells must be induced to proliferate at the expense of nontransformed cells, and this can be achieved by introducing a selectable marker gene and regenerating plants under the appropriate selective regime.

Efficient DNA delivery, competence for regeneration, and a suitable selection system are therefore prerequisites for most plant transformation systems, although there has been recent development in the application of in planta transformation strategies, which circumvent the requirement for extensive tissue culture. Other criteria that define an efficient transformation system are listed in Table elsewhere in this chapter.

9.1.2 DNA Transfer Methods

DNA transfer to plants was first attempted in the 1960s, although the lack of selectable markers and molecular tools to confirm transgene integration and expression made the outcome of such experiments unclear. A breakthrough came in the late 1970s with the elucidation of the mechanism of crown gall formation by *Agrobacterium tumefaciens*.

The discovery that virulent strains of *A. tumefaciens* carried a large plasmid that conferred the ability to induce crown galls and that part of the plasmid (the T -DNA) was transferred to the plant genome of crown gall cells provided a natural gene transfer mechanism that could be exploited for plant transformation.

Tobacco plants carrying recombinant T-DNA sequences were first generated in 1981, although the foreign genes were driven by their own promoters and were not expressed in plant cells. The first transgenic

Table 9.1: Criteria for an efficient plant transformation system.

Essential prerequisites for plant transformation
Efficient method of DNA transfer
Availability of cells/tissues competent for regeneration (not required for in planta transformation strategies)
Suitable selection system
Other criteria that define an efficient transformation system
Reproducibly high transformation efficiency (number of transgenic plants recovered as a proportion of cells/explants originally transformed)
Minimal culture time, to avoid somaclonal variation and sterility.
High-frequency recovery of phenotypically normal, fertile transgenic plants
Technically simple procedure
Economical procedure
Versatile (applicable to many species)
Genotype-independent (applicable to all cultivars and varieties, including elite genotypes)

tobacco plants expressing recombinant genes in integrated T-DNA sequences were reported in 1983.

The technique of *Agrobacterium-mediated* transformation has been developed and refined since then to become a widely used strategy for gene transfer to plants. Although it is convenient and versatile, a major limitation of *Agrobacterium-mediated* transformation is its restricted host range, which until relatively recently excluded most monocotyledonous plants.

The development of strategies to extend the range of plants susceptible to *Agrobacterium* infection is discussed in the following. A number of alternative plant transformation methods were developed to facilitate gene transfer to these recalcitrant species.

These methods can be grouped under the term "direct DNA transfer" and include the transformation of protoplasts using polyethylene glycol (PEG) or electroporation, microinjection, the use of silicon carbide whiskers, and particle bombardment. So far, only direct DNA transfer to protoplasts and particle bombardment have gained widespread use.

The development and application of *Agrobacterium-mediated* transformation, particle bombardment, protoplast transformation, and other transformation techniques is discussed in more detail below.

9.1.3 Cell and Tissue Culture and Plant Regeneration

Small explants of living plant tissue can be maintained on a simple nutrient medium. Transformation may be carried out on tissues dissected from seeds, leaves, stems, roots or buds because under the appropriate conditions these can be induced to dedifferentiate and proliferate to produce undifferentiated callus cultures.

Different hormone treatments induce callus to form shoots and roots, allowing the regeneration of whole plants. Alternatively, the callus can be maintained and subcultured indefinitely or can be broken up in liquid medium to provide a cell suspension culture, which can yield individual cells and protoplasts.

Depending on the transformation method, tissue explants, callus, dispersed cells, or protoplasts can be used as transformation targets. After transformation, cells are allowed to proliferate on selective me dium to increase the amount of callus and kill nontransformed cells.

Transgenic plants can then be regenerated by two methods: somatic embryogenesis or organogenesis. Somatic embryogenesis involves the formation of embryogenic callus direct from somatic tissues. This recapitulates the entire developmental pathway, including the embryonic stage. Organogenesis involves the direct growth of shoots from the callus of transformed tissues or in some cases direct growth from transformed explants without a callus stage.

The shoots can be transferred to rooting medium and regenerated into plants or grafted onto seedling rootstock and propagated. In some species, only one regenerative process is possible under the conditions used for transformation. For example, transgenic rice and maize plants are generated predominantly by somatic embryogenesis, and transgenic cassava plants can be generated only by the organogenesis of shoots.

In other species, such as banana and soybean, both processes are possible and the method of choice depends on the starting material and culture conditions and which process produces transgenic plants the most rapidly and with the greatest efficiency.

An alternative to these processes is the development of transgenic plants by true embryogenesis from transformed seeds, zygotes, or gametes that undergo diploidization. Such techniques do not require extensive tissue culture and are discussed along with in planta transformation strategies later.

9.1.4 Selectable Marker Genes

Most foreign genes introduced into plants do not confer a phenotype that can be conveniently used for selective propagation of transformed cells. For this reason, a selectable marker gene is introduced at the same time as the nonselectable foreign DNA.

This confers upon transformed cells the ability to survive in the presence of a particular chemical, the selective agent, that is toxic to nontransformed cells. In direct DNA transfer methods, the selectable marker and nonselected transgene(s) may be linked on the same cointegrate vector or may be introduced on separate vectors (cotransformation).

Both strategies are suitable because exogenous DNA, whether homogeneous or a mixture of different plasmids, predominantly integrates at a single locus (cointegration). Similarly, separate T DNAs introduced into the same cell tend to integrate at the same locus, although independent integration events have also been demonstrated.

Ideal selection systems should kill nontransformed cells rapidly, but the selective agent (and its detoxification products) should affect neither the physiology nor the regenerative capacity of transformed cells and should be inactivated quickly. The selective regime should be simple and inexpensive to implement.

Most plant selectable marker genes currently in use are dominant selectable markers, i.e., they function in the hemizygous state because they confer a heterologous property upon the plant cell for which there is no competing endogenous activity.

In most cases, selection is toxic to nontransformed cells, but one system confers the ability to use mannose as a carbon source and restricts the growth of nontransformed cells without killing them. Popular selection systems in current use are listed in Table elsewhere in this chapter.

9.2 AGROBACTERIUM-MEDIATED TRANSFORMATION

9.2.1 *Agrobacterium tumefaciens:* A Natural Gene Transfer System

Agrobacterium tumefaciens is a Gram-negative soil bacterium responsible for crown gall disease, a neoplastic disease of many dicotyledonous plants characterized by the appearance of large tumors (galls) on the stems. Virulence is conferred by a large tumor-inducing plasmid (Ti plasmid) containing genes encoding plant hormones (auxins and cytokinins) and enzymes that catalyze the synthesis of amino acid derivatives termed opines.

The plant hormones are responsible for the deregulated cell proliferation that accompanies crown gall growth, while the opines are secreted by the plant cells and used by the bacteria as food. These genes are contained on a specific region of the Ti plasmid, the T -DNA (transfer DNA), so called because it is transferred to the plant nuclear genome under the control of *vir* (virulence) genes carried elsewhere on the Ti plasmid. It is this natural gene transfer mechanism that is exploited for plant transformation.

9.2.2 Development of Ti-Plasmid Vectors

The earliest indication that T-DNA could be used as a plant transformation vector was the demonstration that DNA from an *Escherichia coli* plasmid (the Tn7 transposon) could be stably transferred to the plant genome by first incorporating it into the T -DNA.

However, transgenic plants could not be recovered from the transformed cells, either by regeneration or by grafting onto normal plants, because the hormones encoded by the T-DNA oncogenes caused unregulated and disorganized callus growth. In rare cases, shoots were derived from such callus tissue, and analysis showed that much of the T -DNA (including the oncogenes) had been deleted from the genome.

An important step in the development of T -DNA vectors was the realization that the only requirements for T -DNA transfer to the plant genome were the *vir* genes and the 24-bp direct repeat structures marking the left and right borders of the T-DNA.

No genes within the T -DNA were necessary for transformation, and any sequence could be incorporated therein. This allowed the development of disarmed Ti plasmids lacking all the oncogenes, facilitating T -DNA transfer to plant cells without causing neoplastic growth. Once suitable selectable markers had been incorporated into the T -DNA, Ti plasmids became very powerful gene delivery vectors.

However, wild-type Ti plasmids were unsuitable for the task due to their large size, which made them difficult to manipulate in vitro (large plasmids have a tendency to fragment and they lack unique restriction sites for subcloning).

An early strategy to overcome this problem was the development of intermediate vectors, where the T -DNA was subcloned into a standard *E. coli* plasmid vector, allowing in vitro manipulation by normal procedures, and then integrated into the T-DNA sequence of a disarmed Ti plasmid resident in *A. tumefaciens* by homologous recombination.

This system was simple to use but relied on a complex series of

Table 9.2: Plant selectable marker genes in current use.

Gene (product)	*Source*	*Phenotype and other comments*
aad (aminoglycoside adenyltransferase)	*Shigella flexneri*	Provides resistance to trimethoprim, streptomycin, spectinomycin, and sulfonamides. Used mainly for chloroplast transformation
ble (glycopeptide-binding protein)	*Streptalloteichus hindustantus*	Resistance to the glycopeptide antibiotics bleomycin and pheomycin (and the derivative Zeocin)
dhfr (dihydrofolate reductase)	Mouse	Resistance to methotrexate
sul1 (dihydropteroate synthase)	*E. coil*	Resistance to sulfonamides (Asulam)
epsps (enolpyruvylshikimate	*Petunia hybrida*	Resistance to the herbicide glyphosate phosphate synthase)
hpt (hygromycin phosphotransferase)	*Klebsiella*	Resistance to the aminoglycoside antibiotic hygromycin B. Popular selectable marker in rice
manA (mannose-6-phosphate isome-rase, MIP)	*E. coli*	Ability to grow on mannose as sole carbon source
neo/nptll/aphll (neomycin phosphotransfe rase)	*E. coli*	Resistance to the aminoglycoside antibiotics neomycin, kanamycin, and geneticin (G148)
bar (phosphinothricin acetyltransferase)	*Streptomyces hygroscopicus*	Resistance to phosphinothricin (PPT), which is a component of the herbicides bialophos, Basta, and glufosinate

conjugative interactions between *E. coli* and A. *tumefaciens*, requiring three different bacterial strains (triparental matings). However, because the vir genes act in trans to mobilize the T -DNA, it was soon discovered that the use of large natural Ti plasmids was unnecessary.

Intermediate vectors have been largely superseded by binary vectors, in which the vir genes and the T -DNA are cloned on separate plasmids. These can be introduced into *A. tumefaciens* by conjugation with an E. *coli* donor or by freeze-thaw cycles or electroporation. Most contemporary A *g robacte rium*- mediated transformation systems employ binary vectors.

9.2.3 General Protocol for *Agrobacterium*-Mediated Transformation

At first, *Agrobacterium-mediated* transformation was achieved by cocultivating a virulent A. *tumefaciens* strain containing a recombinant Ti plasmid with plant protoplasts and then obtaining callus from the protoplasts from which fertile plants were regenerated.

This strategy was widely replaced by a simpler method in which small discs were punched from the leaves of recipient species and incubated in medium containing A. *tumefaciens* prior to transfer to solid medium. Infection can be promoted under conditions that induce virulence (presence of 10-200 μM acetosyringone or a-hydroxyacetosyringone, acidic pH, and room temperature), although these need to be optimized for different species.

After coculture for several days, the discs are transferred to a medium containing selective agents to eliminate nontransformed plant cells, antibiotics to kill the bacteria, and hormones to induce shoot growth. After a few weeks, shoots develop from transformed callus cells.

These can be removed and transferred to rooting medium or grafted onto seedling rootstock. Most current protocols for the *Agrobacte*rium-mediated transformation of solanaceous plants are variations on the leaf disc theme, although different tissue explants are suitable transformation targets in different species.

Alternative methods are required for the transformation of monocots. Rapidly dividing embryonic cells (e.g., immature embryos or callus induced from scutellar tissue) are required for the transformation of rice and other cereals. These are cocultured with *Agrobacterium* in the presence of acetosyringone.

9.2.4 Recent Advances- Expanding the *Agrobacterium* Host Range

Although versatile and efficient for many plants, *Agrobacterium-mediated* transformation was, until recently, limited to dicots and monocots of the orders Liliales and Arales (which excludes most of the agronomically important cereals). The range of plants amenable to genetic manipulation by direct DNA transfer is limited only by the availability of cells competent for regeneration, and the range of plants amenable to genetic manipulation by *A. tumefaciens* is further restricted to the species with cells that are both competent for regeneration and susceptible to infection by *A. tumefaciens*.

Driven in part by the potential financial benefits of transgenic crop plants, much research effort has been directed toward extending the *Agrobacterium* host range. In some cases careful optimization of transformation conditions is required. For example, the addition of a surfactant to the inoculation medium was responsible for the successful transformation of wheat embryos and callus.

Such optimization can also dramatically increase the efficiency of transformation in susceptible plants. The duration of cocultivation was shown to be important for the transformation of citrange. The use of feeder cells can also significantly increase transformation efficiency; for example, Niu et al. have shown that cocultivation on a tobacco cell feeder layer significantly increases the efficiency of *Agrobacterium* infection of peppermint.

Artificial infiltration of the bacteria into plant tissues using a syringe or by vacuum infiltration can improve transformation efficiencies in tobacco and is routinely used for the transformation of *Arabidopsis*. A further consideration is the fact that A. *tumefaciens* usually infects wounded cells in its natural hosts, being attracted to phenolic compounds such as acetosyringone that induce expression of the *vir* genes.

Infection and T -DNA transfer are therefore stimulated by wounding tissues (e.g., by crushing or cutting) or by treating explants with acetosyringone. Several monocot species have been infected by pretreating the cells with the exudate from wounded dicot plants, such as potato. The transformation of recalcitrant species such as barley and sunflower has been achieved by prewounding embryonic or meristematic tissue, respectively, with metal particles or glass beads, while sonication has facilitated the uptake of *Agrobacterium* by soybean tissue.

The preinduction of virulence genes may circumvent the necessity to induce a wounding response. Infection may be achieved by testing a range of different *Agrobacterium* strains, and hypervirulent strains such as ALG1 in which the Ti plasmid contains vir genes enhanced by mutation have proved especially useful.

Virulence can also be increased by the use of superbinary vector systems wherein the *vir* component carries multiple copies of the virulence genes. Superbinary vector systems such as pTOK have facilitated the transformation of important monocots, including maize and sugarcane.

Finally, some crops are amenable to *Agrobacterium* infection but resist gene transfer in other ways. Comparative studies of T -DNA gene expression in tobacco and maize cells, for example, have suggested that maize inhibits the integration of T -DNA by suppressing gene expression. Other plants defend themselves against T -DNA integration by forming necrotic boundaries. In some species (e.g., grape), such defenses have been overcome by treatment with antioxidants.

Table 9.3: Recent advances in *agrobacterium* gene transfer technology for the transformation of cereals.

Plant	*Agrobacterium strain*	*Vector name*	*Strain/vector properties*	*Target tissue*
Rice	EHA101 LBA4404	pIG121 pTOK233	Normal strain LBA4404 containing superbinary vector pTOK233 performed cells better than hypervirulent strain EHA101 containing the same superbinary vector	Immature embryos, callus, suspension
Maize	LBA4404	pTOK233	Superbinary vector	Immature embryos
Barley	AGL1	pDM805	Hypervirulent strain	Immature embryo explants
Wheat	C58 (ABI)	pMON 18365	Binary vector	Immature embryos, embryogenic callus

9.2.5 Recent Advances—Increasing the Capacity of T -DNA

Another early problem with *Agrobacterium-mediated* transformation was the inability to transfer large DNA fragments to the plant genome. Although a maximum size limit for T -DNA gene transfer was not determined, the transfer of foreign DNAs larger than 30 kbp was not routinely achievable until about five years ago.

This prevented the transfer of large genes or T -DNA sequences containing multiple genes. The problem was addressed by Hamilton and colleagues, who developed a novel binary vector system in which the T -DNA vector was based on a bacterial artificial chromosome and the *vir* component expressed higher levels of *virG* and *virE*. This bibac (binary BAC) vector facilitates the transfer of up to 150 kbp of DNA to the plant genome, allowing the introduction of gene libraries into plants and the simultaneous introduction of many genes on a cointegrate vector.

9.2.6 Conjugation Systems and Ri Plasmids

The mechanism of T -DNA transfer has many similarities with bacterial conjugation systems, and indeed it has been shown that broad host range bacterial plasmids will transfer from *Agrobacterium to* the plant genome under the control of their own mobilization genes.

Another system with close similarities to the A. *tumefaciens* Ti plasmid is the Ri plasmid of *Agrobacterium rhizogenes*. *This* causes hairy root disease though the integration of its own T -DNA, carrying genes that activate endogenous plant hormones.

Ri plasmids have also been exploited as gene transfer vectors and have the advantage that they do not induce cell proliferation in the host plant (i.e., they are naturally disarmed). *Agrobacterium* strains containing both Ti and Ri plasmids often transfer both T DNAs to the host plant.

The Ri T -DNA induces hairy root disease and acts as a marker for transformation, while the disarmed Ti plasmid carries the transgenes of interest. In this system, dominant markers are not required and regeneration is not prevented by the Ri T -DNA.

9.3 PARTICLE BOMBARDMENT

In 1987, a novel method for plant transformation was introduced by John Sanford and colleagues. This was based on a device that used gunpowder to accelerate small tungsten particles to a velocity of approximately 400 m s^{-1}. The particles were coated with DNA and could penetrate plant cells without killing them.

Initial experiments showed that this device was capable of delivering tobacco mosaic virus (TMV) RNA into onion epidermis cells (resulting in virus coat protein synthesis) and similarly DNA comprising the CaMV 35S promoter and the *cat* reporter gene could be delivered, resulting in high levels of transient CAT activity.

Later experiments showed that particle bombardment was an efficient method for stable integrative transformation. A different device was developed by Agracetus Inc., which used a high-voltage electrical discharge to generate the accelerating force (Accell TM).

The rather erratic gunpowder-based device was refined to a system based on high-pressure blasts of helium, and this is the only commercially available particle gun, which is marketed by Bio-Rad.

9.3.1 Advantages and Disadvantages of Particle Bombardment

Particle bombardment is simple both conceptually and in practice. Typically, plasmid DNA is prepared by standard methods and precipitated onto tungsten or gold particles using $CaCl_2$. Spermidine and PEG are included to protect the DNA during precipitation, and the particles are washed and suspended in ethanol before drying onto Mylar aluminized foil.

This is fired against a retaining screen that allows the microprojectiles through, to strike the target tissue. Particle bombardment is widely used because it circumvents two major limitations of the *Agrobacterium* system. First, it is possible to achieve the transformation of any species and cell type by this method because DNA delivery is controlled entirely by physical rather than biological parameters.

The range of plant species transformable by particle bombardment is therefore restricted only by the competence of cells for regeneration, and the technique is genotype independent and thus useful for the transformation of elite cultivars as well as model varieties. However, careful optimization is required to tailor the method for different species and different cell types and to achieve the highest efficiency transformation with the least cell damage.

Important parameters include acceleration method, particle velocity (controlled by the discharge voltage and/or gas pressure), particle size, and the use of different materials (tungsten, gold). It has also been shown that, for some species and/or tissues, osmotic pretreatment prior to bombardment increases the transformation efficiency.

This can be achieved by partial drying or the addition of osmoticum (mannitol and/or sorbitol) to the culture medium. Optimization experiments are usually carried out by bombarding explants with a screenable marker gene such as *gusA* and assaying for transient expression.

In general, stable transformation by any direct DNA transfer method occurs at a much lower frequency than transient transformation. A number of different reporter genes are used for transient expression analysis in plants as shown in Table elsewhere in this chapter.

Second, particle bombardment allows the stable and heritable introduction of many different genes at once using different plasmids, as these tend to concatemerize to form one DNA cluster that integrates at a single locus. Conversely, multiple transformation using the *Agrobacterium* system requires the cointegration of all the genes in the same T -DNA. Chen et al. reported the cotransformation of rice with 14 separate plasmids containing various marker genes and showed data confirming the cointegration of at least 13 of the plasmids in one plant.

Cotransformation has also been used to introduce up to four agronomically important genes into rice plants, producing plants showing resistance to a spectrum of insect pests and plants with pyramiding resistance against individual pests.

Cointegration prevents different transgenes segregating at meiosis. This can be very important in breeding programs where plants carry two or more transgenes required to generate a single protein, e.g., in plants expressing recombinant human antibodies.

A potential disadvantage of the particle bombardment method is the cost of purchasing or hiring the bombardment device. However, a number of articles have been published providing instructions for the construction of alternative economical devices, such as the particle inflow gun based on flowing helium.

Another disadvantage of particle bombardment is the tendency for DNA sequences introduced by this method to undergo complex rearrangements prior to or during integration. Transgene rearrangement is a pitfall of all direct DNA transfer methods but is perhaps more acute in particle bombardment because the forces involved may cause more DNA fragmentation than other methods and because bombarded plant cells may be induced to produce DNA degradation and repair enzymes in response to their injury.

This may limit the usefulness of particle bombardment for the introduction of large DNA molecules. However, although an upper size limit has not been established, it has been possible to introduce YAC complementary DNA (cDNA) clones into plant cells by this method.

9.3.2 Recent Advances

Recent advances in particle bombardment technology include the clean DNA and agrolistic systems for limiting the amount of plasmid backbone sequence that enters the plant genome, as discussed in more detail below. Successful nuclear integration requires that the metal particle actually enters the nucleus, and recent experiments have shown that transgene integration may be facilitated by damage caused to DNA strands as the particle moves through the nucleus.

Occasionally, exogenous DNA integrates at two different sites separated by megabase pairs of DNA (although these still segregate as a single locus). Confocal analysis of the interphase nucleus following fluorescence in situ hybridization (FISH) to transgene sequences has

Table 9.4: Reporter genes used in plants.

Reporter gene (product)	*Comments*
gusA (β-glucuronidase)	Source: *E. coil* Activity: catalyzes the hydrolysis of p-glucuronides Assays: nonisotropic; in vitro assays are colourimetric or fluorometric; also histochemical assay format using X-gluc Advantages: simple, sensitive, quantitative, many assay formats available, inexpensive Disadvantages: assays are destructive; enzyme is stable so unsuitable for studies of down-regulation
cat (chioramphenicol acetyltransferase)	Source: *E. coli* Tn9 Activity: catalyzes the transfer of acetyl groups from acetyl coenzyme A to chioramphenicol Assays: in vitro assays only, isotropic Advantages: simple to perform Disadvantages: low sensitivity, expensive, low resolution in vivo, reliance on isotopic assay format
luc (luciferase)	Source: The firefiy *Photinus pyralis* Activity: light produced in the presence of luciferase, its substrate luciferin, oxygen, Mg^{2+}, and ATP Assays: nonisotopic bioluminescent assays in vitro and in vivo Advantages: sensitive, rapid turnover, quantitative Disadvantages: expensive detection equipment, limited reproducibility of some assay formats
Anthocyanin regulators	Source: *Zea mays* Activity: induces pigmentation Assays: visual screening for pigmented cells in vivo

(Table Contd.)

	Advantages: simple, inexpensive, nondestructive Disadvantages: low sensitivity, not quantitative, background expression, adverse effects on transgenic plants
GFP (green fluorescent protein)	Source: the jellyfish *Aequorea victoria* Activity: Intrinsic fluorescence under blue-UV light Assays: nonisotopic, in vivo assays in live plants Advantages: intrinsic activity (no substrate requirements), sensitivity, use in live plants Disadvantages: weak signal in some systems (this is being addressed through the use of modifed GFPs with stronger emission and emission at different wavelengths)

shown that these sites may occupy the same region of the nucleus at interphase and that such integration patterns may reflect localized damage to DNA caused by the particles (author's unpublished data, in preparation).

9.4 TRANSFORMATION OF PROTOPLASTS

Protoplasts are plant cells from which the cell wall has been enzymatically or mechanically removed. As with animal cells, the contents of the protoplast cytoplasm are enclosed in a cell membrane, and the transformation of protoplasts can be achieved using many of the procedures routinely used to transfect cultured animal cells.

Two procedures that are commonly used to introduce DNA into protoplasts include the uptake of naked DNA mediated by polyethylene glycol and a divalent cation (either Ca^{2+} or Mg^{2+}) and electroporation, although other agents such as lipofectin have also been used. Plant protoplasts can also be transformed using *Agrobacterium*.

Following transformation, protoplasts are placed on selective medium and allowed to regenerate new cell walls. The cells then proliferate and form a callus from which embryos or shoots and roots can be regenerated with appropriate hormonal treatments.

In principle, protoplasts from any plant species can be transformed, but the technology is limited by the ability of protoplasts to regenerate into whole plants, which is not possible for all species. Although economical and potentially a very powerful procedure, direct transformation of protoplasts is disadvantageous because of the long culture times involved.

Not only does this mean that the transformation process itself is time consuming, but cells cultured for extensive periods either fail to regenerate or frequently regenerate plants that show full or partial sterility and other phenotypic abnormalities (somaclonal variation).

Protoplast preparation, maintenance, and transformation is a skilled technique, which, compared with *Agrobacterium*mediated transformation and particle bombardment procedures, requires a greater investment in training. Advances in other transformation techniques are making protoplast transformation obsolete.

9.5 OTHER DIRECT TRANSFORMATION METHODS

Although the three transformation methods just discussed provide a route for the transformation of most plant species, there is value in the continuing exploration of novel transformation techniques. In some cases, it is possible to learn lessons from animal cell technology, where a number of alternative transformation systems have been established.

Microinjection of DNA into animal eggs and zygotes, for example, is a routine procedure for generating transgenic animals that is applicable to all animal species. Microinjecting DNA into plant cells, while laborious and technically challenging, is advantageous in that it is the only current strategy available to study transformation events at a single-cell level.

The injection of DNA into plant zygotes is being explored as a method for regenerating whole transgenic plants; however, it is currently highly inefficient. For example, isolated maize zygotes microinjected with DNA mimic embryonic development but abort at an early stage. Recently, the same technique has been applied to barley protoplasts, resulting in the recovery of numerous embryo-like structures but only two transgenic plants.

In 1986, Kurata et al. devised a novel transformation strategy for animal cells in which DNA was taken up from solution after piercing the cell membrane with a finely focused laser beam. Although applicable to only a small number of cells, this technique achieved stable transformation efficiencies of 0.5%. A similar technique for plant transformation is available, although stable transformation has not been achieved. Electroporation is used for the transformation of plant protoplasts but has been adapted for the transformation of walled plant cells in tissues.

For example, maize explants have been transformed by electroporation either following partial enzymatic removal of the cell wall or without treatment. This technique has also been successfully applied to rice and sugarcane explants. Pollen has also been transformed using electroporation.

Finally, a novel approach for the transformation of maize cells is to mix cells with silicon carbide whiskers. These penetrate the cells, allowing DNA uptake presumably through transient pores. This method is cheap, reproducible, and very simple to perform. Although initially used only for maize transformation, the technique has now been applied to other cereals.

9.6 TRANSFORMATION WITHOUT TISSUE CULTURE

Most of the transformation procedures discussed so far have one major disadvantage-they rely on tissue culture for the regeneration of whole plants. As discussed, the tissue culture requirement adds significantly to the time and cost of producing transgenic plants and is perhaps the most serious limitation to the technology because it restricts the range of species amenable to genetic manipulation.

Methods that circumvent tissue culture all together -in planta transformation systems-are therefore highly desirable, although until recently they were available only for the model dicot *Arabidopsis thaliana*. These techniques include seed transformation, in planta DNA inoculation, and flower dipping or infiltration, and all involve *Agrobacterium-mediated* transformation.

For seed transformation, *Arabidopsis* seeds are imbibed and cocultivated with *Agrobacterium tumefaciens* and then germinated. This is convenient in *Arabidopsis* because a large number of seeds are produced (10,000 per plant) and the seeds are very small. The procedure is relatively inefficient and is not very reproducible, but because the plant is small it is relatively easy to screen for transformants.

A more efficient procedure is to inoculate plants with bacteria after severing the apical shoots. The bacteria infect the wounded meristematic tissue and about 5% of shoots emerging from the wound site are transgenic. Another method is to dip *Arabidopsis* flowers into bacterial suspension, or vacuum infiltrate the bacteria into the flowers, at around the time of fertilization. This generates plants with transformed and wild-type sectors of vegetative tissue, and these give rise to transgenic progeny at a low frequency. Again, the low transformation frequency is acceptable because of the large number of seeds and the ease of screening.

In planta transformation techniques involving direct DNA transfer have also been developed. These techniques have been applied to a number of economically important crop plants, although they have yet to catch on because they have low reproducibility. Generally, these techniques fall into four classes:

1. Introduction of DNA into meristematic tissue, e.g., by electroporation, injection of DNA/lipofectin complexes, or particle bombardment, followed by the recovery of transgenic shoots-this has been attempted in rice.
2. Introduction of DNA into flowers around the time of fertilization, followed by the recovery of transgenic seeds-this has been attempted in cotton and rice.
3. Imbibing seeds with a DNA solution-this has been used for soybean.
4. Introducing DNA into pollen or mixing pollen with DNA before applying to stigmata-this has been attempted with rice and tobacco.

These techniques have met with variable levels of success.

9.7 PLANT VIRUSES AS GENE TRANSFER VECTORS

Plant viruses do not integrate into the genome and therefore have little value as integrative transformation vectors. They also do not pass through the gametes, so their effects are not heritable through meiosis, although they can be transmitted to new stock by grafting. However, they have a number of attractive properties that suggest they could be developed as episomal gene transfer vectors.

First, they naturally infect intact plant cells including those of species that have, so far, resisted other transformation methods. Viruses exist that infect all the commercially important crop plants, and in many cases even the naked nucleic acid is infectious if applied to, e.g., the leaf surface.

Gene transfer by viral transduction is therefore much easier than *Agrobacterium-mediated* transformation, much simpler than protoplast transformation, and much cheaper than particle bombardment. Second, because they multiply to high copy numbers in the plant cell, viruses can facilitate very high-level transgene expression. Third, many plant viruses cause systemic infections, i.e., they spread to every cell in the host plant, therefore obviating the need to regenerate transgenic plants from transformed cells.

These properties have led to the development of a number of RNA virus-based expression vectors for recombinant protein synthesis, including those based on tobacco mosaic virus, potato virus X, and cowpea mosaic virus. Plant viruses can also be used as a sensitive assay to detect T -DNA transfer to the plant nucleus.

Since viruses replicate to such high copy numbers, viral infection symptoms can be used to confirm the presence of T-DNA in single cells of a given explant. This procedure, termed agroinfection, involves the transfer of a viral genome into the plant cell between T-DNA border repeats and was first used to confirm the genetic transformation of maize.

9.8 ORGANELLE TRANSFORMATION

So far, only transformation of the nuclear genome has been considered. It can be desirable, however, to introduce DNA into organelle genomes as this increases the number of transgene copies in the cell and facilitates highlevel transgene expression.

The first reports of organelle transformation were serendipitous. For example, de Block et al. reported the transformation of tobacco chloroplast DNA using *Agrobacterium-mediated* transformation of tobacco protoplasts.

The rare organellar transformation event was discovered because of the maternal transmission of the transgene, confirmed by Southern blot analysis of chloroplast DNA. The targeted transformation of chloroplasts by particle bombardment was achieved in 1988 by Boynton and colleagues using the unicellular green alga *Chlamydomonas rheinhardtii*.

The same technique was used by Svab and Maliga to transform tobacco chloroplast DNA. In each case, the particles penetrated the double membrane of the chloroplast so that a biological mechanism to pass through the membrane was unnecessary.

More recently, PEGmediated transformation of protoplasts has been used to introduce DNA into the tobacco chloroplast genome and the two methods have been compared. Generally, only one of the many organelles in the cell is transformed, which means that homoplastomic cells must be obtained by intense selection.

In most reports, the *aad* selectable marker gene has been used. A recent strategy to track chloroplast transformation involves the fusion of this selectable marker to the gene for green fluorescent protein.

9.9 FUTURE PROSPECTS-CONTROLLED TRANSGENE INTEGRATION AND EXPRESSION

9.9.1 Overview

While the major goal of plant transformation technology over the last two decades has been to introduce and stably express transgenes in plants, the challenge for the next decade is to refine these techniques and introduce single transgene copies at defined sites without extraneous DNA sequences such as parts of the vector backbone or marker genes.

The ultimate aim would be to produce transgenic plants with the transgene integrated at a known site, allowing the experimenter full and predictable control over transgene expression.

9.9.2 Clean DNA Transformation

A major disadvantage of all direct DNA transfer transformation methods is the tendency for exogenous DNA to undergo rearrangement and recombination events leading to the integration of multiple fragmented, chimeric, and rearranged transgene copies.

Such clusters are prone to internal recombination and may therefore be unstable, as well as promoting transgene silencing. Kohli et al. have shown that such rearrangements may be caused, in part, by recombinogenic elements in the vector backbone of the transformation plasmid.

The backbone also contains bacterial markers and other sequences, such as the origin or replication, which are unnecessary for transformation. A major breakthrough has been achieved in the last year in the production of transgenic plants without integrated vector backbone sequences.

Fu et al. describe a clean DNA technique involving the use of linear minimal transgene cassettes for particle bombardment. These cassettes lack all vector backbone elements but contain the DNA sequences essential for transgene expression: the promoter, open reading frame, and terminator.

Transgenic rice plants regenerated from callus bombarded with such cassettes showed very simple integration patterns, with one or a few hybridizing bands on Southern blots. Comparative transformation experiments using whole plasmid DNA resulted in much more complex integration patterns, with multiple bands of different sizes.

Furthermore, release of a diagnostic fragment from the transgene showed that fewer rearrangement events had occurred in the plants carrying the linear cassettes. The progeny of these plants were examined for transgene expression, and there were no reported instances of transgene silencing.

Conversely, whole plasmid transformation by particle bombardment routinely leads to silencing in about 20% of the resulting transgenic plants. This work also showed that linear cassettes promoted the same efficiency of cotransformation as normal plasmids, so the clean DNA strategy is a promising technique for the generation of transgenic plants carrying multiple transgenes.

Agrobacterium-mediated transformation was for a long time regarded as a naturally clean system, in which one to three intact T -DNA copies integrated into the genome, and only the DNA between the T -DNA border repeats was transferred. More recent data show, however, that DNA outside the border repeats is cotransferred to the plant genome at a very high frequency (>75% of transformants).

In many such cases, the backbone is linked to the T -DNA, suggesting that the VirD2 protein can skip over one T -DNA border, whereas in other cases the backbone segments are not linked to the T -DNA, suggesting an adventitious integration mechanism.

Furthermore, T -DNA often integrates as multiple copies, predominantly inverted repeat structures, which are highly correlated with transgene silencing phenomena. Multimeric integration also occurs when two different T DNAs are transferred to the same plant cell by different *Agrobacterium* strains, suggesting that multimer formation occurs before integration.

This study showed a preference for linkage involving the right border repeat, although all other possible linkage combinations occurred at a lower frequency. Complex T -DNA integration patterns are more prevalent with hypervirulent *Agrobacterium* strains and/or superbinary vectors. To address such problems, Chilton and colleagues have pioneered the technique of agrolistic transformation.

Particle bombardment was used to introduce three plasmids into tobacco and maize cells, one containing a recombinant T -DNA and the others containing the *virDi* and *virD2* genes responsible for T -DNA mobilization. The resulting plants contain single integrated T -DNA sequences at high frequency, indicating that the Vir proteins could function correctly, although T -DNA sequences were also detected as part of the entire plasmid integrated into the plant genome.

So, although this technique is useful for the generation of clean integration events, it does not eliminate backbone sequences completely. A possible strategy to limit the production of vector-containing plants is to flank the T -DNA with "killer genes" that confer a lethal phenotype on the transformant.

Any plants carrying vector backbone sequences would be killed, provided the lethal genes were not disrupted or mutated upon integration and that segments of the vector backbone distal to the lethal genes did not integrate. A more refined technique devised by Gelvin is to produce single-strand T -DNA in vitro using highly purified recombinant VirD2 protein.

This development, analogous to the clean DNA strategy because backbone sequences are never introduced into the plant, may allow genuinely clean T-DNA insertions. However, the T -strands must be introduced into protoplasts by electroporation; thus, the procedure is conceptually no different from direct DNA transfer.

9.9.3 Toward Marker-Free Transgenic Plants

There is some debate in the public domain about the potential dangers of selectable marker genes in transgenic plants. The permanent integration and expression of a selectable marker also prevents the same plant line from being retransformed with the same marker (i.e., if it was necessary to introduce further transgenes into an existing transgenic plant line).

This issue has been tackled in a number of ways:

(1) the use of genetic systems to eliminate marker genes from transgenic plants,

(2) the development of strategies for marker gene segregation in breeding lines, and

(3) the development of innocuous markers that do not confer antibiotic or herbicide resistance.

The first genetic system for marker elimination was reported by Goldsborough et al. and involved the incorporation of transgenes within a maize Ac or *Ds* transposon, cloned within the T -DNA sequence.

Two marker genes were used, and plants could be recovered that lacked one or other of the markers. In a subsequent system, the *Agrobacterium* isopentenyl transferase *(ipt)* gene was cloned in an Ac element within the T -DNA borders, while a second reporter gene carried in the T -DNA was outside the transposon.

The *ipt* gene affects cytokinin metabolism so transformed plants showed a shooty phenotype. *Ac* elements often jump from site to site in the genome, but in about 15% of cases there is excision without subsequent reintegration. In these cases, the marker was removed and normal plant growth resumed as long as further copies of the T -DNA were not present.

The excision events left the remainder of the T -DNA intact and did not affect expression of the linked reporter gene. This system is advantageous for a number of reasons:

(1) the marker is innocuous,
(2) the marker can be eliminated, and
(3) there is automatic selection for single-copy insertions of the T -DNA.

In sexual populations, one method that could produce marker-free transgenic plants is to use cotransformation to introduce the nonselected gene and selectable marker at different loci and screen progeny in the R1 generation for segregants lacking the selectable marker gene.

In direct DNA transfer cotransformation, there are occasional dispersive integration events resulting in transgene segregation and aberrant segregation ratios, although cointegration at a single locus predominates.

The majority of T-DNA cotransformation events also result in cointegration, so that only a low frequency of marker-free plants is generated. The use of superbinary vector systems with two T -DNA plasmids may facilitate marker segregation, and segregation has been shown in 50% of the progeny of rapeseed and tobacco plants produced by *Agrobacterium-mediated* cotransformation with two binary plasmids.

The preceding strategy cannot be used for vegetatively propagated plants. An alternative strategy utilizes the bacteriophage site-specific recombinase system *Cre-loxP*. Tobacco plants were transformed with two marker genes, each flanked by *loxP* recombination target sites.

The plants were then retransformed with the *cre* gene, and expression of Cre recombinase resulted in recombination between the *loxP* sites and excision of the original markers.

One problem with this system is that transformation with the *cre* gene must itself involve an additional selectable marker. This requirement was circumvented by the transient expression of Cre, resulting in the production of a few marker-free plants.

9.9.4 Toward Precise Integration and Control of Expression

Another way in which *Cre-loxP*, and other site-specific recombination systems, can be exploited is to target transgenes to precise loci in the plant genome.

Even in the absence of multiple and rearranged transgene copies, there is still much variation in transgene expression levels, and this has been attributed to so-called position effects reflecting the nature of the local molecular environment at the site of integration (e.g., the presence of regulatory elements, repetitive DNA, chromatin structure, and DNA methylation).

In animal cells, a number of loci have been identified that are favourable to transgene expression because transgenes integrated there tend to be expressed faithfully and strongly. The controlled integration of DNA at such sites can be facilitated by flanking the transgene with *loxP* sites and placing *a loxP* site at the desired target site.

A similar approach has been attempted in plants, but it appears that transgenes integrated at sites on several different chromosomes still display variable levels of expression within lines of identical transformants, suggesting that other factors are responsible for the variation.

This agrees with data showing variable levels of transgene expression and different modes of transgene silencing involving different methylation patterns in a line of rice plants carrying the same single-copy, three-gene transgenic locus.

A different strategy to avoid position effects is to isolate the transgene from the surrounding molecular environment by flanking it with matrix attachment regions (MARs) that attach it to the nuclear matrix and isolate the transgene into an independent chromatin domain.

The variability in transgene expression was reduced when reporter genes such as gusA were flanked by MARs, although as with the *Cre-loxP* system, there was still some variability that could be attributed to other factors.

9.10 CONCLUDING REMARKS

It is now possible, at least in principle, to introduce any gene into any plant species. Problems of efficiency, cost, intellectual property, and other constraints remain but these are not insurmountable.

The next few years will see further refinements in transformation technology, but it is likely that the focus will shift from technology development and optimization of the DNA transfer process itself to more challenging and fundamental territories, such as elucidating the factors that permit stable and predictable transgene expression, efficient transformation of organelles, and homologous recombination.

Pathway engineering, involving the simultaneous introduction and concerted expression of multiple transgenes, will have to feature prominently in future strategies for plant transformation.

Finally, the outputs from current genomics, proteomics, and metabolomics programs can be fully exploited only with sophisticated methods to introduce large pieces of DNA into the plant genome.

10

Indole Alkaloid Production by Genetic Engineering

Metabolism is the closely coordinated series of enzyme-mediated reactions in a living organism such as a plant. This complex biochemical network consists of metabolic pathways in which compounds that have direct importance for the vital functions of an organism are synthesized and utilized.

This collection of processes is defined as primary metabolism and the compounds involved are primary metabolites, e.g., sugars, amino acids, fatty acids, nucleotides, and the polymers derived from them (polysaccharides, proteins, lipids, DNA, RNA, etc.).

Furthermore, in all plants more specialized biochemical pathways exist, known as secondary metabolism, in which a wide range of so-called secondary metabolites are produced, e.g., alkaloids, anthocyanins, flavonoids, quinones, lignans, steroids, and terpenoids.

Secondary metabolites play a role in the interaction of the plant with its environment, such as attraction of pollinating insects via colour and scent and protection against predation by herbivores and insects or against infection by microorganisms. These secondary metabolite pathways are often restricted to an individual species or genus and might be activated only during particular stages of growth and development or during periods of stress caused by, e.g., wounding, attack by microorganisms, or limitation of nutrients.

Many intermediates are utilized in both primary and secondary metabolism. Therefore, the dividing line between primary and secondary metabolism is not always distinct. These overlapping roles of intermediates cause a close interaction between primary and secondary metabolism.

Therefore, the regulatory mechanisms in plants that provide control at the cellular, genetic, protein, and molecular levels of secondary metabolism are the subject of considerable research. The major interest in plant secondary metabolites is due to their functions for the plant, which are often connected with their impressive biological activities and are the basis for commercial applications such as pharmaceuticals, insecticides, flavors, colourants, nutrition, spices, or fragrances.

Because of various difficulties involved in growing plants in the field or collecting them in the wild (geographic location, seasonal variation, pests and diseases, risk of species extinction, political instability of the country where the plantations are, etc.), researchers are focusing on plant cell and tissue cultures as a production system as this system can be used under controlled conditions for the production of secondary metabolites of commercial interest.

It is widely recognized that cultured plant cells represent a potential source off commercially valuable plant metabolites. Some successes in production of secondary metabolites by plant cell cultures have been achieved, e.g., shikonin production by cultures of *Lithospermum erythrorhizon* Siebold & Zucc. (Boraginaceae), berberine production by cultures of *Coptis japonica (L.)* Salisb. (Ranunculaceae), and ginsenoside production by cultures of *Panax ginseng C.A.* Mey. (Araliaceae).

Taxol is the latest accomplishment with an increase of 10to 20-fold in productivity compared with the average *Taxus* (Taxaceae) cultures. However, there are some major obstacles to the production of secondary metabolites by plant cell cultures: the spectrum of compounds produced is quite different from that in the plant, the yield of the product(s) for economically viable production is too low, and they suffer from heterogeneity and/or instability during long-term culture.

As in many cases productivity is too low for commercialization, various strategies to improve productivity are being followed:

- Cell line improvement: screening and selection for high-producing cell lines
- Optimization of growth and production media (phytohormones, nutrients, precursors, antimetabolites)
- Optimization of culture conditions (inoculum size, agitation, temperature, light, pH, shaker speed, immobilization, permeabilization, two-phase systems, two-stage systems, elicitation)

- Cultures of differentiated cells (in vitro shoot or root cultures, hairy roots)
- Metabolic engineering

Although the production of differentiated cells has resulted in improved productivity of the desired compounds (e.g., for tropane alkaloids), for large-scale industrial production the requirement for very special bioreactors is a major economic constraint.

Therefore, there is great interest in the last mentioned approach, metabolic engineering, the more so as it has the potential for application to plant cell cultures, microorganisms, and the plant. Also, it has the possibility of application to isolation of enzymes.

Further applications of metabolic engineering of secondary metabolism are new compounds with biological activity, new flower or food colours, new taste or fragrance of food, improved nutritional effect of food, decreased levels of undesired compounds in food and fodder, and improved resistance against pests and diseases.

For increasing the production of desired compounds, different strategies can be considered: increasing the carbon flux through a biosynthetic 'pathway by affecting, e.g., rate-limiting enzymes, feedback inhibition, and competitive pathways; increasing the number of producing cells; and decreasing catabolism.

Besides the previous efforts using random mutation and selection, current genetic engineering offers the tool for a focused approach to improve secondary metabolite production along these lines.

10.1 CATHARANTHUS ROSEUS AS A SOURCE OF TERPENOID INDOLE ALKALOID PRODUCTION

Catharanthus roseus plant cell cultures are a potential commercial source for the production of some pharmaceutically important alkaloids. *Catharanthus roseus (L.)* G. Don is a tropical plant belonging to the family Apocynaceae and produces a wide range of terpenoid indole alkaloids (TIAs). More than 130 alkaloids have been isolated from different parts of the plant and many of them show biological activities, such as the antineoplastic vinblastine and vincristine, the antihypertensive ajmalicine, and the sedative serpentine.

Cell suspension cultures of C. *roseus* do not represent an exception for secondary metabolite production regarding the obstacles mentioned before. They produce a much simpler alkaloid profile, the yield is less than that from intact plants, and after prolonged subculture they lose the

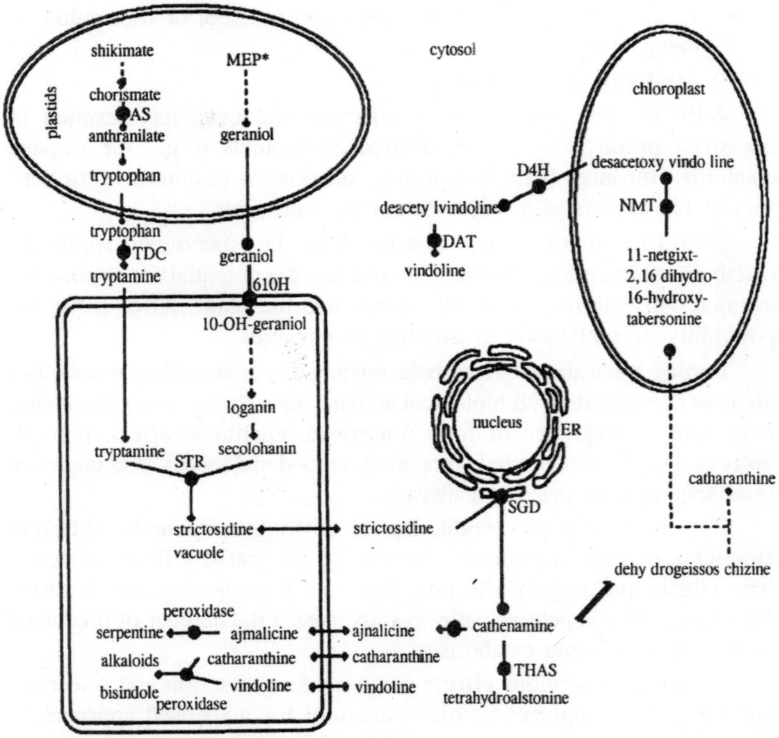

Figure 10.1: Subcellular compartmentation of terpenoid indole alkaloid biosynthesis in *Catharanthus roseus*.

ability to produce alkaloids at high levels even after extensive screening and selection procedures.

Therefore, the regulation of the biosynthesis of TIAs in cell cultures of *C. roseus* is being extensively studied in our laboratory in a collaborative project, the BSDL-project group Plant Cell Biotechnology, with the Institute of Molecular Plant Sciences Leiden, and the Department of Biochemical Engineering of the Delft University of Technology.

The TIA pathway is a complex metabolic network. The biosynthetic precursors for TIA biosynthesis are provided by the shikimate pathway and the terpenoid pathway. It was proved that the terpenoid moiety of the TIM, secologanin, is not derived from the mevalonate pathway but instead from the MEP pathway using a cell suspension culture of *C. roseus*. Because in the tryptophan branch of the shikimate pathway tryptophan decarboxylase (TDC, EC 4.1.1.28) operates at the interface between primary and secondary metabolism, it has been demonstrated

Table 10.1: Properties of the Enzyme Tryptophan Decarboxylase (TDC, EC 4.1.1.28) from *Catharanthus roseus*.

Localization	Cytosol
Solubility	Soluble
Purification	Apparent homogeneity
Isomers	One
In vivo activity	In vivo equilibrium between an active form and a monomeric form that is susceptible to irreversible inactivation. When the equilibrium is shifted to the monomeric form of TDC, the protein could be degraded in the ubiquitin proteolytic pathway.
Relative M, on PAGE	115 ± 3 kDa
Subunits	Two (54 ± 1 Da), (54 and 56 kDa)
Cofactors	Pyridoxal phosphate (0.5-1 mol), pyrrolo
quinolinequinone PI	5.9
pH optimum	8.5
Stability (half-life time)	For purified TDC at 0°C, pH 7.5 about 4 days, for partially purified TDC 7 days For crude enzyme at -15°C about 7 days
Cooperative effects	
L-Tryptophan	$R_s = 80$, $n_H = 1.06$
Substrate specificity	
Accepted	4-Methyltryptophan, 5-fluorotryptophan, L-tryptophan ($K_m = 7.5 \times 10^5$ M), L-5-hydroxytryptophan ($K_m = 1.3 \times 10^3$ M)
Not accepted	p-Fluorophenylalanine, aminoethylcysteine, 5-methyltryptophan, L-phenylalanine ($K_m = 8$), L-tyrosine ($K_m = 8$), L-dopa ($K_m = 8$).
Inhibitors	Tryptamine (competitive, K_I 3.1×10^{-4} M), o-tryptophan (noncompetitive), carbon dioxide (noninhibitor)

to be a site for regulatory control of TIA biosynthesis. TDC catalyzes the conversion of L-tryptophan to tryptamine, the indole moiety of the TIA. Strictosidine synthase (STR) catalyzes the condensation of tryptamine with secologanin by a reaction of the Pictet-Spengler type into 3α(S)-

strictosidine, the pivotal intermediate of the TIM. The properties of TDC and STR from *C. roseus* are listed in Tables elsewhere in this chapter, respectively. A number of enzymes of the TIA biosynthetic pathway have been characterized and the encoding genes have been cloned.

This has enabled studies of the overexpression of these genes in various plants and organisms. There are some valuable examples of what metabolic engineering can achieve in secondary metabolism using some of these genes.

The seeds of canola *(Brassica napus)* are used as animal feed, but the presence of indole glucosinolates makes the crop less palatable. After the transformation of B. *napus* with the *Tdc* gene from *C. roseus*, the mature seeds of the transgenic plants contain reduced levels of indole glucosinolates, which increases the economic value of the plant.

Another example is the functional expression of the *Str* gene from *Rauwolfia serpentina* in bacteria, yeast, and insect cells, which can be used as a tool for production of large amounts of STR. Another good example is transgenic yeast cultures containing the *Sir* and/or *Sgd* (strictosidine glucosidase) genes that can be grown on the juice of *Symphoricarpus* berries, which contains both sugar and secologanin, and are fed tryptamine.

The system produces large amounts of mostly strictosidine ($2gL^{-1}$) and cathenamine. However, overexpression of the *Tdc* or *Sir* genes does not always lead to the desired results. For example, expression in *Tabernaemontana pandacaqui* did not affect production of TIM, and in transgenic *Peganum harmala* hairy roots overexpressing the *Tdc* gene the production of the desired harman-type alkaloids did not occur due to the conversion of tryptamine to serotonin.

The results of (over)expressing the enzymes TDC, STR, and TDC and STR in *C. roseus* and other species are shown in more detail in Tables elsewhere in this chapter, respectively.

10.1.1 Aspects of TIA Formation by Transgenic Cell Lines of *C. roseus* Overexpressing *Tdc* and *Str*

The highest ajmalicine production ever achieved in our laboratories was 77 mg L', obtained after 16 days in a bioreactor under optimized conditions with a wild-type cell line of *C. roseus*. Such productivity is too low for commercialization; therefore, extensive studies were made to characterize genetically modified *C. roseus* cell cultures, in particular with regard to TIA formation, to investigate the effect of overexpression

Table 10.2: Studies with (Over)expression of the Tryptophan Decarboxylase (Tdc) Gene

Tdc	
Origin of the clone	*Catharanthus roseus (L.)* G. Don (Apocynaceae)
Promoter	lac a
Expressed in	*Escherichia coil*
Results	No attempt was made to quantify the level of the TDC activity in *E. coil.*
Tdc	
Origin of the clone	*C. roseus (L.) G.* Don (Apocynaceae)
Method	*Agrobacterium-mediated* transformation using A. *tumefaciens*
Promoter	Cauliflower mosaic virus (CaMV) 35S
Expressed in	*Nicotiana tabacum* 'Xanthi' (Solanaceae)
Obtained as	Transgenic plants
Results	Leaves of transformed plants had 4 to 45 times more TDC activity than the controls. Increase in tryptamine accumulation: >1 mg g^{-1} FW plant with the highest TDC activity, a 260-fold increase over control The growth and development of the transformed plants appeared normal.
Tdc	
Origin of the clone	*C. roseus (L.) G.* Don (Apocynaceae)
Method	Agrobacterium-mediated transformation using A. *tumefaciens*

(Table Contd.)

	Promoter	CaMV 35S
	Expressed in	N. tabacum'Xanthi' (Solanaceae)
	Obtained as	Transgenic plants
	Results	TDC activity in the transformed tobacco was 45 times greater than in the controls and resulted in a >200-told increase in the free tryptamine and >30-fold in free tyramine pools.
Tdc		
	Origin of the clone	*C. roseus (L.) G.* Don (Apocynaceae)
	Method	*Agrobacterium-mediated* transformation
	Promoter	CaMV 35S
	Expressed in	*Peganum harmala L.* (Zygophyllaceae)
	Obtained as	Cell suspension and hairy root cultures
	Results	Despite increased TDC activity, tryptamine level did not increase due to its conversion into serotonin. Consequently, serotonin level increased, but no increase in desired harman-type alkaloids was observed.
Tdc		
	Origin of the clone	*C. roseus (L) G.* Don (Apocynaceae)
	Method	Agrobacterium-mediated transformation using A. *tumefaciens*
	Promoter	CaMV 35S
	Expressed in	N. tabacum 'Petit Havana' SR1 (Solanaceae) Obtained as Plants
	Results	The plants with highest TDC activity contained 19 pkat/mg protein.

(Table Contd.)

	Specific activity of AS and CM was not affected by expression of TDC.
	Despite increased tryptamine accumulation, the phenotype of the transgenic plants' was normal.
Tdc	
Origin of the clone	*C. roseus (L.) G.* Don (Apocynaceae)
Method	*Agrobacterium-mediated* transformation using A. *tumefaciens*
Promoter	CaMV 35S
Expressed in	*Brassica napus* 'Westar' (Brassicaceae)
Obtained as	Transgenic plants
Results	TDC activity in transgenic plants varied and was 4 to 45 times greater than in contrtols
	Transgenic plants accumulated tryptamine while correspondingly lower levels of tryptophan-derived indole glucosinolates were produced in all plant parts compared with controls.
	Indole glucosinolate content of mature seeds from transgenic plants was only 3% of that found in nontransformed seeds.
Tdc	
Origin of the clone	*C. roseus (L.) G.* Don (Apocynaceae)
Method	*Agrobacterium-mediated* transformation using A. *tumefaciens*
Promoter	CaMV 35S
Expressed in	C. roseus'Moming Mist' (Apocynaceae)

(Table Contd.)

	Observed as	Crown-gall calli
	Results	An increase in TDC protein level, TDC activity, and tryptamine content but no significant increase in TIA production were observed In *Tdc*-antisense-containing calli decreased TDC activity was measured
Tdc		
	Origin of the clone	*C. roseus (L.) G.* Don (Apocynaceae)
	Method	*Agrobactenum-mediated* transformation using A. *tumefaciens*
	Promoter	CaMV 35S
	Expressed in	*Solanum tuberosum* 'Desiree' (Solanaceae)
	Observed as	Transgenic plants
	Results	Phenylpropanoid pathway can be down-regulated by the control of substrate supply created by the introduction of a single gene outside the pathway. The redirection of tryptophan to tryptamine resulted in a decrease in the levels of tryptophan, phenylalanine, and phenylalanine-derived phenolic compounds in transgenic tubers. Chlorogenic acid was 2- to 3-fold lower in transgenic tubers. Transgenic tuber became more susceptible to infection with *Phytophthora unfastens* due to alterations in the substrate levels in the shikimate pathway and phenylpropanoid pathways.
Tdc		
	Origin of the clone	*C. roseus (L.) G.* Don (Apocynaceae)

(Table Contd.)

Method	*Agrobacterium-mediated* transformation using A. *tumefaciens*
Promoter	CAMV 35S
Expressed in	*N. tabacum* SR1 (Solanaceae)
Observed as	Transgenic plants

of the *Tdc* and *Str* genes on TIA formation and to follow their stability during long-term subculture.

The available knowledge on wild-type cell cultures of *C. roseus* was the basis for these further studies with transgenic cell lines. Transgenic cell lines were obtained from *C. roseus* seedling leaf cells by means of *Agrobacterium* tumefaciens-mediated transformation. These cells overexpress the genes *Str (S* lines) and/or *Tdc* (T lines).

The enzymes STR and TDC catalyze key steps in the biosynthesis of TIAs. All our experiments were carried out using a two-stage production system in which biomass and product formation occur sequentially. Sufficient biomass was first obtained in growth medium (GM), and the cells were then inoculated into production medium (PM).

Our PM does not contain phosphate, nitrate, or phytohormones, constituents that promote culture growth. Very little cell division therefore occurs in PM; most biomass increase is the result of starch biosynthesis. As shown in Table elsewhere in this chapter, after the initiation of the transgenic calli lines we obtained 10 transgenic cell lines with six different T -DNA constructs.

These transgenic cell cultures showed wide phenotypic diversity for TIA accumulation: non-TIA-accumulating and TIA-accumulating cell lines. After the initiation of transgenic cell suspension cultures from these calli lines, several cell lines gradually lost their viability after a number of subcultures. The surviving productive cell lines S 10, S 1 (S lines, transgenic for STR), and T22 (T line, transgenic for TDC) were chosen for further studies of TIA accumulation. Both S and T lines showed higher STR and TDC activities, respectively, than the wild-type levels.

The TIA accumulation profiles of the productive cell lines were as follows: line S10 produced ajmalicine, serpentine, catharanthine, and tabersonine and lines S1 and T22 produced strictosidine, ajmalicine, serpentine, and catharanthine. Most notably, the viable transgenic cell lines showed many years of transgene stability.

The transgenic cell lines were initiated from seedling leaf cells by means of *A. tumefaciens*. Therefore, these cell lines might arise from one or just a few transformed cells, which carries the risk that effects of the transformation are related to the site of the insertion of the foreign genes.

The S lines seemed more productive than the T lines after the initiation of the transgenic cultures. Overexpression of the enzyme TDC increased the tryptamine production in most of the T lines compared

Table 10.3: Studies with (Over)expression of the Strictosidine Synthase (Str) Gene

Str		
	Origin of the clone	*Rauvolfia serpentina (L.)* Kurz (Apocynaceae)
	Method	Baculovirus-based expression
	Promoter	Baculovirus polyhedrin *(polh)*
	Expressed in	*E. coli*, *Saccharomyces cerevisiae*, *Spodoptera frugiperda* (fall arm worm)
	Results	The clone has been identified. *Str* has been expressed in an enzymatically active form in *E. coil.* Addition of equimolar (15 mM) concentrations of secologanin and tryptamine to the bacteria in either exponential or stationary phase resulted in the quantitative conversion of the precursors to strictosidine. Enzyme was in a soluble form within the bacterium but strictosidine accumulated in the medium. Expression of Str in S. *frugiperda* resulted in the overproduction of STR in a catalytically active form (max. 4 mg L^{1})
Str		
	Origin of the clone	*C. roseus (L.) G.* Don (Apocynaceae)
	Method	*Agrobacterium-mediated* transformation using A. *tumefaciens*
	Promoter	CAMV 35S
	Expressed in	*N. tabacum* 'Xanthi' (Solanaceae)
	Observed as	Transgenic plants

(Table Contd.)

	Results	The transgenic plants had 3 to 22 times greater STR activity than C. *roseus* plants. One difference in expression of STR between tobacco and C. *roseus* was the presence of a lower molecular weight in tobacco. Two distinct forms of the enzyme were produced in transgenic plants but only a single form was made in C. *roseus* plants. The second form of the protein is not a result of overexpression in tobacco but may represent differences in protein processing between tobacco and C. *roseus*.
Str		
	Origin of the clone	C. roseus (L.) G. Don (Apocynaceae)
	Method	Expression cassette polymerase chain reaction
	Promoter	*tac*
	Expressed in	*E. coil*
	Results	Induction of the *Str* gene resulted in overexpression of the STR, which accumulated mainly as insoluble inclusion bodies. Denaturation and refolding of the insoluble protein resulted in the ability to purify up to 6 mg of active enzyme from a single liter of cell culture. The recombinant STR activity was about 30 nkatlmg.
Str		
	Origin of the clone	*C. roseus (L.) G.* Don (Apocynaceae)
	Method	Agrobacterium-mediated transformation using A. *tumefaciens*
	Promoter	CAMV 35S

(Table Contd.)

	Expressed in	*Tabemaemontana pandacaqui* Poir. (Apocynaceae)
	Observed as	Transgenic plants
Str		
	Origin of the clone	*Rauvoiia serpentina* Poir. (Apocynaceae)
	Expressed in	22 bacterial strains *(Aeromonas* sp., *Brochothrix thermosphacta, B. campestris, Bacillus lincheniformis, Cellulomonas uda, Citrobacter freundii, Enterobacter aerogenes, E. cloacae, Escherichia coil, E. coli T, Kiebsiella oxytoca, K. pneumoniae, K. terrigena,* Listerta *grayi, L. innocus, Proteus mirabilis, P vulgaris, Salmonella typhimurium, Serratia marcescens, Staphylococcus aureus, S. epidermidis, S. camosus)*
	Results	Strictosidine was converted by the bacterial strains into vallesiachotamine and isovallesiachotamine. *Str* and *Sgd*
	Origin of the clone	*C. roseus (L.) G. Don* (Apocynaceae)
	Method	Lithium acetate procedure
	Expressed in	*Saccharomyces cerevisiae* (yeast)
	Results	With secologanin and tryptamine feeding, transgenic yeast that was introduced with the genes *Strand* Sgd (strictosidine glucosidase) produced 2 g L^{-1} strictosidine. Strictosidine is excreted into the medium and only small amounts of cathenamine were found. Growing the transgenic yeast on the sap obtained by pressing *Symphoricarpus* berries, which contain both sugar and secologanin, and feeding tryptamine, the system was able to produce large amounts of strictosidine.

Table 10.4: Studies with (Over)expression of the *Tdc* and *Str* Genes.

Tdc and *Str (74)*	
Origin of the clone	*C. roseus (L.) G.* Don (Apocynaceae)
Method	Agrobacterium-mediated transformation using *A. rhizogenes*
Promoter	CaMV 35S
Expressed in	*Nicotiana tabacum* 'Petit Havana' SR1 (Solanaceae), *Lonicera xylosterum L.* (Caprifoliaceae), *Weigela* 'Styriaca' (Diervillaceae), *C. pusillus'Murray) G.* Don (Apocynaceae), *Cinchona* officinalis'Ledgeriana' (Rubiaceae) Observed as Transgenic hairy root cultures
Results	In tobacco TDC and STR activities and tryptamine were detected.
	High levels of secologanin (up to 1.25 mg g-' FW) was produced by *L. xylosterum* cultures.
	The *Weigela* cultures after about 2 months turned brown and stopped growing.
	Different TDC and STR activities between two hairy root cultures of *C. pusillus* were observed.
	TDC and STR activities up to 4.1 and 450 pkat mg^{-1} of protein, respectively, were recorded. High levels of tryptamine and strictosidine, up to 1.2 and 1.95 mg g^{-1} DW, respectively, were measured.
Tdc and Str (75)	
Origin of the clone	C. roseus (L.) G. Don (Apocynaceae)
Method	Agrobacterium-mediated transformation using A. *tumefaciens*

(Table Contd.)

Promoter	CAMV 35S
Expressed in	*Nicotiana* tabacum'Petit Havana' SRI (Solanaceae)
Observed as	Transgenic cell suspension cultures
Results	Transgenic tobacco cells showed relatively constant TDC and 2- to 6-fold greater STR activities compared with the non transformed tobacco cell in which the enzyme activities were not detectable. The transgenic culture accumulated tryptamine and produced strictosidine upon feeding of secologanin.
Tdc and *Str (76)*	
Origin of the clone	*C. roseus (L.) G.* Don (Apocynaceae)
Method	*Agrobacterium-mediated* transformation using A. *rhizogenes*
Promoter	CaMV 35S
Expressed in	*Cinchona officinalis* 'Ledgeriana' (Rubiaceae)
Observed as	Transgenic hairy root culture
Results	The products of TDC and STR, tryptamine and strictosidine were found in high amounts, 1.2 and 1.9 mg m^{-1} DW. However, no increase in desired quinoline alkaloids was observed compared with the amounts of 0.05 mg g^{-1} FW reported by Hamill et al.. One year later they lost the capacity to accumulate alkaloids.
Tdc **and** *Str (78)*	
Origin of the clone	*C. roseus (L.) G.* Don (Apocynaceae) *(Tdc*; *Str)*
Method	Particle bombardment

(Table Contd.)

Promoter	CaMV 35S
Expressed in	*Nicotiana tabacum* 'Samsun' NN
Observed as	Transgenic plants
Results	In 26% of the 150 independent transgenic plants, both the plant *Tdc* and *Stri* transgenes were silenced, 41% demonstrated a preferential silencing of either transgene, 33% of them expressing both transgenes. Seven-week-old seedlings showed 24- and 110-fold variation in levels of TDC and STR1 activities, respectively.
Tdc and *Str* (Hilliou et al., personal communication)	
Origin of the clone	*C. roseus (L.) G. Don* (Apocynaceae)
Method	Particle bombardment
Promoter	CaMV 35S
Expressed in	*C. roseus (L)* G. Don (Apocynaceae)
Observed as	Transgenic calli cultures
Results	Despite an increase in TDC and STR activities observed compared with the nontransformed cultures, TIA accumulation did not increase.

with the wild type. When S lines had detectable levels of TDC activities, they were more productive, improving the flux through the pathway before the terpenoid pathway became a limiting factor for TIA formation during long-term subculture.

10.1.2 Effect of Culture Conditions on the Productivity of the Transgenic Cell Lines

Accumulation of TIA was strongly influenced by culture conditions, such as the hormonal composition of the medium and the availability of precursors. Alkaloid accumulation by highly productive transgenic lines showed significant instability in the first year of their maintenance.

Table 10.5: Gene constructs and derived cultures.

	Number of established callus lines			
T-DNA constructs	*Total*	*Antibiotic resistant*	*GUS positive*	*Established cell cultures*
<npt-gus>-Str>-Tdcl>	38	1	1	ST1
<hpt-<Tdc2-<Str-gus>	146	8	5	None
<npt-<gus-<Str	45	2	2	S1
<hpt-<gus<Str	330	22	19	S4, S6, S7, S10, S11
<npt-gus>-Tdcl>	47	2	2	T11
<hpt-gus>-Tdcl>	288	11	9	T21
<hpt-<Tdc2-<gus	50	2	2	T22

One of the most productive transgenic cell lines, S 10, was used as a model to study the effect of the presence of the synthetic auxins naphthalene acetic acid (NAA) and 2,4-dichlorophenoxy acetic acid (2,4-D) in the culture medium on the accumulation of TIAs.

It was found that the presence of NAA during the production phase led to lower levels of alkaloid accumulation, whereas the presence of 2,4-D in the growth medium significantly reduced culture aggregation and measurably repressed the TIA biosynthesis due to reduced capacity to supply biosynthetic precursors.

At the enzyme level, the expression of the reporter gene β-glucuronidase *(Gus)* and the *Str* gene was not affected by auxins under the conditions tested. As can be seen in Fig elsewhere in this chapter, the productivity of these cell lines increased in the sequence GM < PM < PMa < PMb.

10.1.3 Precursor Feeding Experiments: Identification of Bottlenecks and Determination of the Capacity of the Transgenic Cell Lines for TIA Accumulation

As overexpression of the genes *Tdc* and *Str* did not result in a much increased TIA level during long-term subculture, the total capacity of the pathway was assessed by feeding the indole and iridoid precursors. The uptake and utilization rates of the fed precursors differed among cell lines.

Lines S 10, S1, and T22 and the wild-type cell line CRPM were fed in PM with various concentrations and combinations of the indole precursors L-tryptophan and tryptamine and the iridoid precursors loganin and secologanin. First, the optimal conditions for feeding were determined.

It was found that addition of precursor(s) in the early culture and to the PM was more efficient than addition to the GM. From feeding a single precursor, it was learned that addition of the indole precursors such as L-tryptophan and tryptamine to the culture medium did not increase TIA production but addition of the iridoids such as loganin and secologanin did increase alkaloid production.

This shows that the major bottleneck for the biosynthesis of TIAs was in the terpenoid pathway for both the transgenic lines and the wild-type cell line CRPM. Interestingly, feeding with loganin, which first need to be converted to secologanin, was more efficient than feeding with secologanin.

It is thought that exogenous secologanin is trapped in an inactive pool in a different compartment; e.g., secologanin can react with free amino groups of proteins. Feeding tryptamine in combination with loganin was also more efficient than the L-tryptophan/loganin combination.

This suggests that even in the *Tdc*-overexpressing cell line the conversion of L-tryptophan to tryptamine remains a rate-limiting step for *TIA* formation. This can be overcome by addition of tryptamine to the culture medium, which results in an increase of *TIA* accumulation.

Despite this limitation in the biosynthetic rate of tryptamine, lines S1 and T22 could supply endogenous tryptamine as much as the 0.8 mM and 1.6 mM loganin-fed cultures, respectively. In line S10 after feeding, no accumulation of tryptamine is observed; instead, it is rapidly converted via strictosidine into *TIAs*.

In lines S1 and T22 strictosidine was accumulated. Considering possible regulation aspects of the feeding, it can be concluded that loganin feeding seems to increase production of *TIAs* by mobilizing tryptamine. This can be either by inducing tryptamine production or by

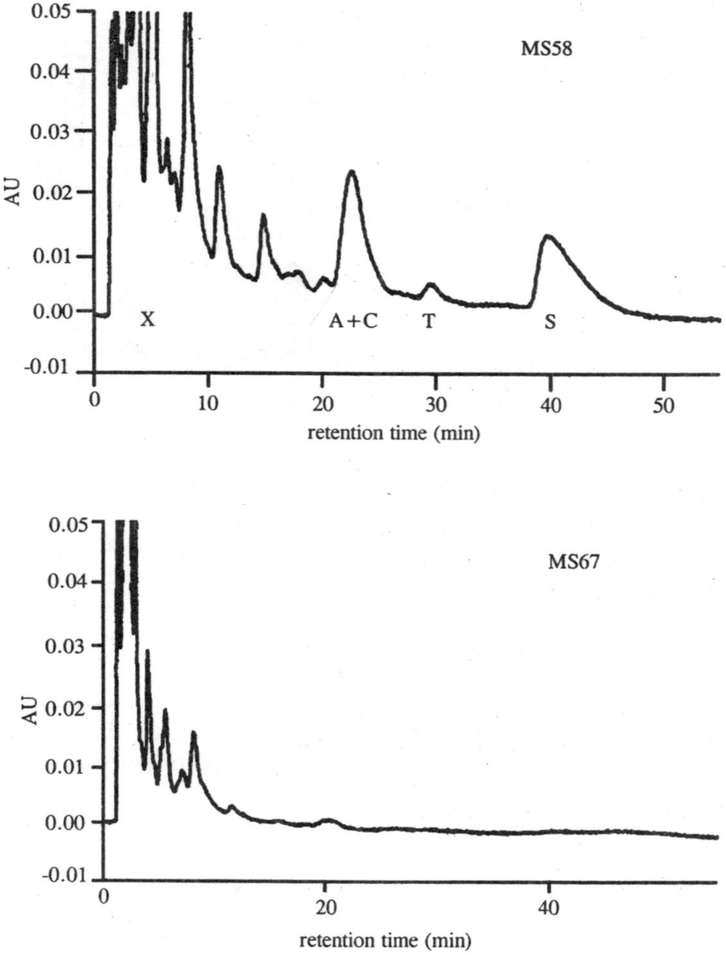

Figure 10.2: HPLC chromatogram showing the effect of growing S10 cells in MS58 [per liter: Murashige and Skoog salts, 0.1 g myo-inositol, 0.4 mg thiamine, 2 mg NAA, 0.2 mg kinetin, and 30 g sucrose] and MS67 (per liter: MS salts, 0.1 g myo-inositol, 1 mg thiamine, 0.5 mg pyridoxine, 0.5 mg niacin, 2 mg glycine, 1 mg 2,4-D, 0.2 mg kinetin, and 30 g sucrose) on alkaloid accumulation. Absorbance at 280 nm by compounds eluding from μBondapak Phenyl column (Waters Chromatography, Etten-Leur, The Netherlands) at a flow rate of 2 mL min^{-1}. The identity of the compounds was established by photodiode array detection of their absorption spectra. X, unidentified nonalkaloid compounds; A, ajmalicine; C; catharanthine; S, serpentine; T, tabersonine.

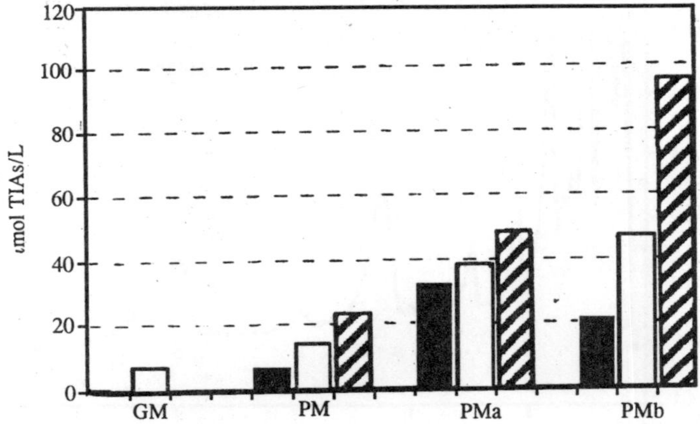

Figure 10.3: Terpenoid indole alkaloid accumulation by transgenic cell suspension cultures of S10 (black bars), S1 (dashed bars) (both transgenic for *Str*), and T22 (white bars) (transgenic for *Tdc*) in different media [GM; PM; PMa; PMb: PMb differ from PMa with 80 g L^{-1} sugar content instead of 50 g L^{-1}]. Duplicate cultures were pooled upon harvesting on day 14 and analyzed as single samples.

channeling tryptamine away from a competitive pathway into *TIA* biosynthesis. On the other hand, addition of L-tryptophan or tryptamine does not seem to induce either loganin or secologanin availability.

Under optimized experimental conditions, cultures of line S 10 accumulated 350 μmol L^{-1}, line SI about 600 μcool L^{-1}, and line T22 1200 μmol L^{-1} of TIAs while the wild-type cell line accumulated about 580 μcool L^{-1} (manuscript in preparation). The profile of TIAs was cell line specific, and the time courses of accumulation of individual TIAs differed.

The accumulation of large amounts of TIAs by the non-TIA-accumulating wildtype cell line CRPM after feeding tryptamine and loganin suggests that the cells have an enormous capacity to produce TIAs. In all experiments, catabolism of the alkaloids was observed. Very little is known about the products formed in catabolism.

10.1.4 Stability of the Transgenic Cell Lines over a Period of 30 Months of Subculture

The transgenic cell lines were not selected for high productivity, but they were maintained in medium containing antibiotic, either hygromycin or kanamycin, for transgene selection and gave GUS positive results. The STR and TDC activities showed fluctuations, although they remained significantly higher than in the wild-type line.

However, production of TIAs decreased from several mg L^{-1} to an

Table 10.6: Summary of TIA accumulation by some cell lines in production medium (PM).

	Fed and harvest (day)	*Type of feedings*	*TIAs*			
			Fed	*Nonfed*	*TDC*	*STR*
S10	FOH3	0.5L	183	76	ND	ND
	F11 H14	0.5 L	212	194	ND	ND
	F11 H14	0.5 LT	352	194	ND	ND
S1	FOH2	0.8-6.4 L	~250	50	*	*
	FOH2	1.6 LT	~700	50	5	229
	FOH2	1 L	250	38	5	114
	FOH 14	1 L	600	110	6	381
	FOH2	1 LT	471	38	22	129
	FOH 14	1 LT	515	110	11	344
T22	FOH2	1.6-6.4 L	~510	10	*	*
	FOH2	6.4 LT	1100	10	90	89
	FOH2	1 L	250	—	38	71
	FOH14	1 L	48	2	57	141
	FOH2	1 *LT*	300	—	57	49
	FOH 14	1 *LT*	39	2	117	81
	FO-10 H2	0.4 LT	290	—	8	45
	FO-10 H8	0.4 LT	1250	19	7	35
CRPMQ	FOH2	1 LT	114	—	ND	ND
	FOH14	1 LT	587	—	ND	ND

undetectable level during long-term subculture. The decrease in the production of TIAs is thought to be the result of selection of fast-growing cell types, which in terms of secondary metabolism are not active or have new bottlenecks occurring in the secondary metabolite biosynthetic pathways (manuscript in preparation).

10.1.4 Conclusions and Prospects

Biosynthesis of TIAs involves different cellular compartments and even different cells for some of the final steps in biosynthesis of the dimers, although it has been shown for anthocyanins in *C. roseus* cell cultures that the level of accumulation in the culture is determined by the percentage of producing cells. It is also not known whether only a small part of the cells is producing alkaloids. Apparently, over expression of *Str* in combination with feeding tryptamine and loganin did not

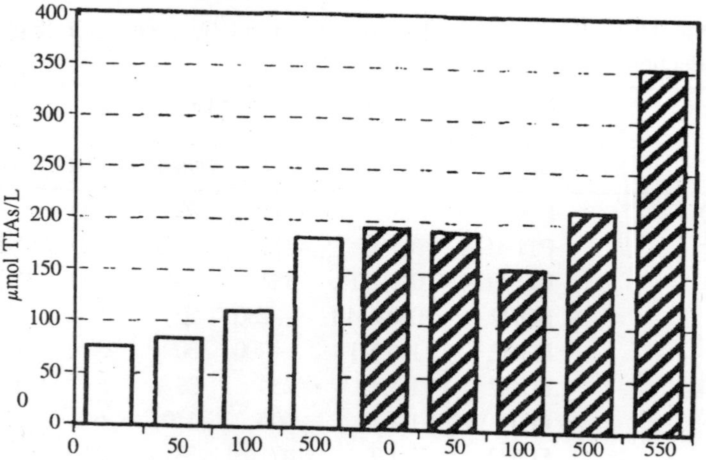

Figure 10.4: Effect of feeding loganin on TIA accumulation by line S10 (transgenic for *Str)* under conditions of low tryptamine availability. White bars represent culture treated on day 0 and sampled on day 3; hatched bars represent culture treated on day 11 and harvested on day 14. The concentration of exogenously added precursors is expressed in μM. C indicates control; Land LT indicate that loganin and both loganin and tryptamine were added, respectively. Triplicate cultures were pooled upon harvesting and analyzed as single samples.

lead to increased alkaloid production compared with feeding of wild-type cells.

This might be explained by having some physiological barriers for alkaloid production. The availability of tryptamine in producing cells might be mentioned in this connection. Maybe overexpression of *Tdc* overcomes the problem of availability of this precursor in the producing cells as it results in a much higher alkaloid production capacity after feeding precursors.

Transgenic cell lines with increased TIA accumulation lost this trait after prolonged subculture. Apparently, instability still appears to be a problem as it is for nontransgenic cells. Constitutively increased enzyme activity in the TIA pathway, in this case TDC or STR, was not sufficient to keep elevated TIA levels.

Although TDC is known to be a strongly regulated enzyme, which has a transient maximum in activity shortly after subculture on a fresh medium, TDC seems to not be a limiting factor in these cells. Also, STR is not a limiting factor. The following points might be considered in future studies, concerning possible physiological or further biochemical barriers for TIA accumulation in *C. roseus* cell cultures:

The cellular or subcellular localization of enzymes, precursors,

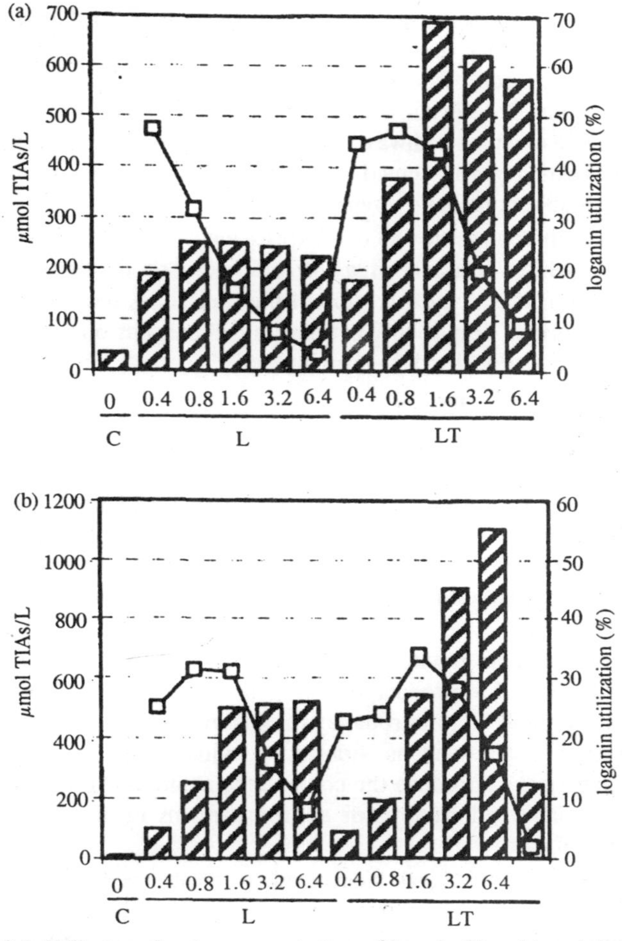

Figure 10.5: Utilization of various concentrations of loganin (L) and loganin/tryptamine (LT) by lines S1 (transgenic for *Str*, (a) and T22 (transgenic for *Tdc*, *(b)*. Feeding consisted of addition of L or LT in a series of concentrations from 0.4 mM up to 6.4 mM at inoculation. Right-hand y-axis represents percentages of loganin (o) utilized in TIM. Duplicate cultures were pooled upon harvesting on day 2 and analyzed as single samples (C, control).

and TIAs

Factors involved in transport of the precursors for TIAs Effect of competitive pathway(s) utilizing TIA precursors Catabolism of TIAs

Factors that control the carbon flux into the pathway

As all the factors are connected with each other, further studies on the regulatory genes and the overexpression of such genes seem a promising approach. Our studies showed that there is a short-term increase in the TIA levels with overexpression of *Tdc* and *Str* genes; however, in a complex pathway such as the TIA pathway, overexpression of one or two enzymes seemed to be sufficient only when accompanied by a favourable epigenetic environment.

This is in contrast to the overexpression of chalcone isomerase in tomatoes, which resulted in a 70-fold increase of flavanoids. An encouraging fact is that expression of the transgenes seems to be stable over long periods despite the fact that in cell suspension cultures there is a strong selection for fast-growing cells.

10.2 THE FUTURE

Since about 6000 BC, when the Sumerians and Babylonians in Mesopotamia produced alcoholic beverages, cheese, and other products, biotechnology has been part of our life and it will only increase in importance. Now, we are at a stage where plant and plant cell culture technology, which has taken big steps since Haberlandt's predictions in 1902 about cultivation of plant tissue, is being combined with genetic engineering.

Plants such as potatoes, alfalfa sprouts, and tomatoes can be genetically altered to produce the antigens for hepatitis B and cholera. The next step will be to introduce proteins involved in the biosynthesis of pharmaceuticals such as vinblastine, vincristine, or taxol in the appropriate host to produce the compounds mentioned on an industrial scale with plant cell cultures or microorganisms in bioreactors or by transgenic plants.

To accomplish this goal, biosynthetic pathways need to be completely identified on gene, enzyme, and product levels. In particular, identification of factors involved in the regulation of the biosynthetic pathways seems an interesting approach that may lead to regulatory genes controlling the flux through a pathway. This will make possible the production of known or novel fine chemicals by plant cell cultures.

11

Seed Proteins by Genetic Engineering

The protein content of seeds varies widely, from about 10-12% in cereals to 20-25% in pulses and oilseeds and up to 50% in soybean. It can be calculated that the total protein yield of the major seed crops amounts to over 350 million tonnes per annum, with cereals accounting for about 70% of the total.

Seeds therefore form the major source of protein for the nutrition of humans and livestock. In addition, seed proteins are largely responsible for the processing properties of many seeds including the use of wheat for making bread and other processed foods such as pasta and noodles. The impact of seed proteins on the nutritional and processing quality of seeds has provided a major stimulus to their study over a period exceeding 250 years since Beccari first described the isolation of wheat gluten.

Seed proteins also provided an attractive model system to study gene structure and expression, as their synthesis is tightly regulated with high levels occurring in specific seed tissues and at specific stages of development.

Consequently, complementary DNAs (cDNAs) and genes for seed proteins were among the earliest to be cloned from plants, providing a basis for longer term strategic studies of the structures and functional (i.e., processing) properties of the encoded proteins. This has led over the past few years to success in improving aspects of seed quality by genetic engineering, although no improved varieties have so far reached the level of commercial production.

Furthermore, the capacity of seeds to synthesize and store high levels of protein, combined with well-established systems for seed production, harvesting, storage, and fractionation, has led to the use of seeds as bioreactors to produce novel proteins and peptides, notably high value products for biomedical applications. These form the subject of other chapters in this volume and therefore, focus on the manipulation of seed proteins to improve their properties for traditional end uses (i.e., food and feed). However, before proceeding with this we briefly discuss the structures of seed proteins and their impact on quality.

11.1 SEED PROTEINS

Seed proteins can be broadly classified into two groups, storage and nonstorage. Storage proteins can be defined as proteins that have no biological role except to provide a store of carbon, nitrogen, and sulfur that is mobilized during germination to support seedling growth.

The amounts of storage proteins present vary between seeds of different species, accounting for about half of the total seed proteins in cereals but a higher proportion in more protein-rich oilseeds and legumes. Furthermore, the amounts also vary with the nutritional status of the plant, with high levels of nutrient availability resulting in increased accumulation.

Storage proteins are invariably deposited in discrete protein bodies, where they form dense masses and do not interfere with essential cellular processes. They usually accumulate in cotyledonary and/or endosperm tissues, although storage in the perisperm or hypocotyl may also occur.

Storage proteins are usually classified into four groups on the basis of their solubility properties and sedimentation coefficients (S_{20w}) following a system and using a nomenclature standardized by Osborne.

11.2 GLOBULIN STORAGE PROTEINS

The most widespread storage proteins are globulins, which are defined by their solubility in dilute salt solutions. These comprise 7-8S and 11-12S types that are often called vicilins and legumes, respectively, reflecting their characteristic presence in legume seeds. The 7-8S globulins are typically trimeric molecules with three subunits of M_r about 40-50,000 forming a protein of M_r about 150-190,000.

However, in some species post-translational proteolysis and/or glycosylation of the subunit chains results in a much more complex spectrum of components. For example, pea contains unprocessed subunits of M_r about 50,000 and a range of smaller components (M_r 12,000, 13,000, 16,000, 19,000, 25,000, 30,000, 33,000, and 34,000), the latter all being derived from M_r 47,000 precursor subunits.

The 78S globulin storage proteins occur in many but not all legumes, in other dicotyle-donous plants such as cotton and cocoa, and in the embryos and aleurone cells of cereals. The 11-12S globulins differ from the 7-8S globulins in being hexameric, typically comprising six subunits of M, about 60,000. Posttranslational proteolysis results in cleavage of the subunits into acidic (also called a or A) chains of M_r about 40,000 and basic (J3, B) chains of M_r about 20,000, which remain associated by a single disulfide bond.

Further proteolysis does not occur and glycosylation is rare. The 11-12S globulins are widely distributed in legumes and other dicotyledonous seeds (e.g., composites, crucifers, cucurbits) with related proteins forming the major storage protein fraction in oats and rice and minor components in wheat. Although they have little sequence homology, the 7-8S and 11-12S globulins have similar three-dimensional structures, which may facilitate packing in the protein bodies, and appear to be derived from a common ancestral protein.

11.3 PROLAMINS

Prolarnins are characterized by their insolubility in water or dilute salt solutions but solubility in alcohol-water mixtures. They are major storage proteins in most cereal species (except oats and rice, where they are minor components) and vary widely in their structures and properties.

The individual components are characterized by high contents of proline and glutamine (from about 30 to 70% of the total residues, respectively) and the presence of repeated amino acid sequences, while some individual subunits may be assembled into oligomers or polymers stabilized by interchain disulfide bonds.

11.4 2S ALBUMINS

These are typically low-M_r proteins derived from single polypeptide chains that are posttranslationally cleaved to give subunits of M_r about 4-5000 and 8-9000 that remain associated by two interchain disulfide bonds. They are widespread in dicotyledonous plants (e.g., legumes, crucifers, composites), where they occur together with 7-8S and/or 11-12S globulins.

11.5 NUTRITIONAL QUALITY OF SEED PROTEINS

Animals are able to synthesize only about half of the 20 amino acids commonly found in proteins, the remainder being required in their diet and hence termed "essential." If only one of these is limiting, the remaining amino acids cannot be utilized but are broken down and excreted, leading to poor feed conversion and environmental pollution. This is not usually a problem with ruminants, in which the rumen bacteria provide all 20 amino acids, but is a problem for nonruminants such as pigs and poultry. Comparison of the amino acid compositions of seeds with the requirements of essential amino acids for humans and nonruminant livestock shows that both legumes and cereals are deficient in one or more amino acids.

Furthermore, these deficiencies are determined by the amino acid compositions of the storage protein fractions. The prolamins of wheat, barley, and maize contain low levels of lysine (=1 mol % or less), and this is the first limiting amino acid in all three species. The prolamins (zeins) of maize also contain little or no tryptophan, which is the second limiting amino acid in this species, and all three also contain low levels of threonine.

In contrast, the 7S and II S globulins of legumes are severely deficient in cysteine and methionine, resulting in deficiencies in these amino acids in the whole seeds. Because cereal and legume seeds are largely complementary in their contents of essential amino acids, they are often used in appropriate mixtures to provide a balanced diet of essential amino acids.

Nevertheless, there is considerable interest in producing high-lysine cereals and high-sulfur legumes to provide more flexibility in producing low-cost diets for livestock.

11.6 HIGH-LYSINE CEREALS

Interest in improving the nutritional quality of cereals was stimulated by the discovery in the 1960s and 1970s of a range of high-lysine mutants in maize, sorghum, and barley. The first of these were the *opaque2 and loury2* mutants of maize, followed by the lys gene of Hiproly barley and ***hl*** mutant of sorghum.

These are all spontaneously occurring mutant genes; a range of other spontaneous high-lysine genes were subsequently identified in maize, and a range of chemical and physical mutagens were used to generate further high-lysine genes in sorghum and barley.

Despite initial optimism, only one of these mutant genes, *opaque2* of maize, has been successfully exploited in plant breeding programs, using genetic modifiers to convert the soft starchy endosperm to a normal type. Although the other mutants have failed to be exploited because of associated deleterious effects on yield, analysis of their composition and molecular basis has contributed to our understanding of the regulation of seed proteinn biosynthesis and allowed us to make more informed decisions on strategies for improving quality.

The high-lysine phenotype of most mutant lines is due, at least in part, to a reduction in the proportion of the lysine-poor prolamins and increases in other more lysine-rich proteins. It is, therefore, tempting to suggest that high-lysine cereals could be engineered by down-regulation of prolamin gene expression using antisense or cosuppression technology, as described for 2S albumins (napins) and globulins in transgenic oilseed rape.

In this work, antisense technology was used to down-regulate either napin or 12S globulin gene expression in developing seeds, resulting in compensatory increases in the remaining protein group. However, in all high-lysine cereals the decreased prolamin content is associated with reduced starch synthesis and hence lower yield, and this has proved difficult to separate from the high-lysine character by plant breeding.

Consequently, reduction in prolamins is not generally considered to be a valid strategy for engineering high-lysine cereals. Instead, three different strategies have been proposed.

11.6.1 Manipulation of the Amino Acid Composition of Prolamins

Wallace et al. designed modified maize prolamins (zeins) containing single and double substitutions of lysine for neutral amino acids; insertions of oligopeptides (five to eight residues) rich in lysine, threonine, and tryptophan; and insertion of an M_r 17,000 peptide derived from the SV40 VP2 protein.

All except for the latter (included as a control) were synthesized and aggregated into dense particles when the messenger RNA (mRNA) was injected into egg cells of the frog *Xenopus,* indicating that the proteins should also be stable in transgenic cereals. However, the drawback of this strategy is that zeins account for about half of the total protein in the maize seed and appear to be encoded by up to 100 genes.

Consequently, it would be necessary to express the mutant zeins under control of a highly active promoter and possibly also to combine this with down-regulation of the endogenous wild-type zeins. A similar strategy could be applied to other cereals.

11.6.2 Expression of Specific Lysine-Rich Proteins

Detailed analyses of the high-lysine barley mutant Hiproly demonstrated that about half of the increase in lysine resulted from elevated levels of four specific lysine-rich proteins, which together account for about 17% of the total grain lysine, compared with about 7% in normal lines.

These proteins are β-amylase (=5.0 g % Lys), protein Z (now known to be a serpin proteinase inhibitor) (=8.1 g % Lys), and chymotrypsin inhibitors CI-1 (=9.5 g % Lys) and CI-2 (=11.5 g % Lys). CI-1 and CI-2 were both purified from Hiproly, providing a basis for molecular cloning.

This demonstrated that both proteins occur in two isoforms with the major form of CI-2 comprising 84 residues with an M_r of 9,380. This form contains seven lysine residues (8.3 mol %) but no cysteine residues (and hence no disulfide bonds). The three-dimensional structure of the protein shows a wedge-shaped disc of about 28 × 27 × 19 Å with a single α-helix of 3.6 turns and four strands of β-sheet with a lefthanded twist.

The reactive (inhibitory) site (Met-59) of CI-2 is present on an exposed nine-residue loop (Gly-54 to Arg-62). Perutz and coworkers have shown that it is possible to mutate or extend this loop region, inserting up to 10 glutamine residues at Met-59 in order to explore the role of glutamine repeats in neurodegenerative diseases. More recently, Roesler and Rao have described the design and expression in *E. coli* of five mutant forms of CI-2 containing 20-25 mol % lysine, based mainly on mutation of surface-exposed residues.

Although all the mutants appeared to have conformations similar to that of the wild-type protein, as determined by circular dichroism spectroscopy, they had lower thermodynamic stabilities. Nevertheless, at least one of the mutants was considered to be suitable for expression in transgenic plants.

The thionins are another group of lysine-rich seed proteins. Rao et al. designed high-lysine analogues of barley α-hordothionine, a 45residue protein containing five lysines (11 mol %). These included an analogue containing seven additional lysines (i.e., a total of 27 mol %), which was synthesized and shown to exhibit antifungal activity similar to that of the wild-type protein.

The authors have not, however, attempted to express any of the mutant proteins in transgenic plants, and such material would perhaps be unlikely to be accepted by consumers or regulatory authorities because of the wide biological activity of thionins.

Pea legumin contains about 5 mol % lysine, which is similar to the nutritional requirements for humans (5.5 g %). Sindhu et al. reported

Table 11.1: The Essential Amino Acid Contents of Soybean, Maize, Wheat, and Barley Seeds and Their Major Storage Protein Fractions Compared with the WHO Recommended Levels.

	Soybean			*Maize*			*Wheat*			*Barley*		
Amino acid	*Total seed*	*7S congly-cinin*	*12S glycinin*	*Total seed*	*Ethanol-soluble zein*	*Ethanol-insoluble zein*	*Total seed*	*Gliadins*	*Glutenins*	*Total seed*	*Hordein*	*WHO*
Cysteine	1.3	0.3	1.4	3.1	1.7	5.0	2.6	3.0	2.4	2.9	1.9	3.5
Methio nine	1.3	0.2	1.8	2.0	1.6	7.9	1.3	1.7	1.4	1.7	1.1	
Lysine	6.4	7.0	4.9	3.5	0	0.1	2.0	0.6	2.3	3.1	0.8	5.5
Isoleucine	4.5	6.4	4.7	3.6	4.4	2.5	3.6	4.3	3.7	3.6	0.4	4.0
Leucine	7.8	10.2	7.2	11.6	23.0	13.9	6.7	6.8	6.6	7.2	7.8	7.0
Phenylalanine	4.9	7.4	5.7	4.9	7.1	3.6	5.1	5.5	4.8	5.5	9.3	6.0
Tyrosine	3.1	3.6	4.1	2.3	6.5	6.3	2.6	2.6	3.6	2.7	4.3	
Threonine	3.9	2.8	3.4	3.9	3.1	3.7	2.7	2.2	3.1	3.3	1.7	4.0
Tryptophan	1.3	0.3	1.6	0.9	nd	nd	1.1	0.7	2.1	2.0	nd	1.0
Valine	4.8	5.1	5.1	4.9	3.7	4.6	3.7	4.1	4.2	4.6	3.5	5.0
Histidine	2.5	1.7	2.2	3.2	1.2	4.8	2.2	2.2	2.3	1.9	1.8	—
Notes	b	c	d	e	f	9	b	9	9	b	h	i

that pea legumin was stably accumulated in the seeds of transgenic rice plants but did not determine the effect on the lysine content of the grain.

An alternative approach to using naturally occurring high-lysine proteins, or mutant forms of these, is to design high-quality proteins ab initio. Keeler et al. demonstrated the feasibility of this concept by designing low-M, proteins (3600 to 6700) with an α-helical coiled coil structure and containing up to 43 mol % lysine accompanied by high levels of methionine and in some cases also tryptophan.

The proteins were initially synthesized in *E. coli* and one protein containing 31 mol % lysine and 20 mol % methionine was then expressed in tobacco seeds under control of the bean phaseolin and soybean 8-conglycinin promoters.

Increases in lysine of up to 0.8 mol %, compared with a wild-type level of 2.56 mol %, were observed in the primary transformants and were inherited through three generations. No impact on seed methionine levels was reported, and the work has not so far been extended to cereals or other crop plants.

11.6.3 Increased Accumulation of Free Lysine

The most spectacular success in increasing the lysine content of cereals has been achieved by increasing the amount of free rather than protein-bound lysine. This is based on eliminating the feedback regulation that normally limits the pool of free lysine by transforming with genes for feedback-insensitive enzymes from microorganisms.

Two enzymes are important in this respect: aspartate kinase (AK), which catalyzes the first reaction in the pathway, and dihydrodipicolinate synthase (DHDPS), which controls the branch point leading to lysine. Falco et al. initially showed that a twofold increase in total lysine occurred in seeds of oilseed rape (canola) expressing a DHDPS gene from *Corynebacterium* and a fivefold increase occurred in soybean seeds expressing the same enzyme together with an AK gene from *E. coli*.

More recent work, described in a patent application, has shown that expression of the DHDPS gene in maize under control of the embryo + aleurone-specific globulin-1 promoter resulted in increases in free lysine from 1.4 to 15-27% of total free amino acids, representing increases in total lysine from 2.3 to *3.6-5.3%*. *In* contrast, no effect was observed when the same gene was expressed under control of the starchy endosperm-specific glutelin-2 promoter, which was thought to be due to increased lysine catabolism.

It was concluded that down-regulation of lysine catabolism would be required in order to achieve accumulation of lysine in the maize starchy endosperm.

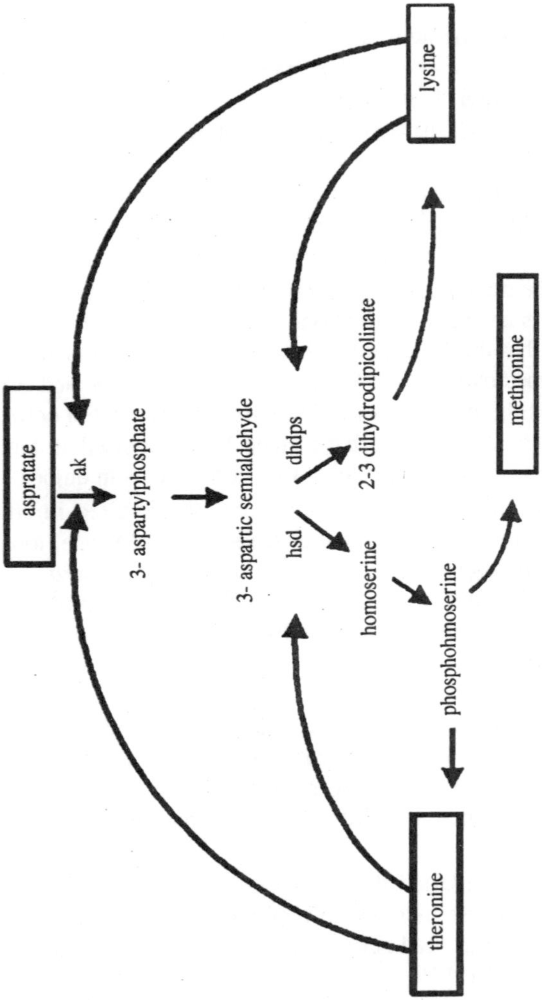

Figure 11.1: The pathway of lysine biosynthesis in plants showing feedback regulatory steps as arrows. AK, aspartate kinase; DHDPS, dihydrodipicolinate synthase.

11.7 HIGH-METHIONINE LEGUMES

Legume seeds are deficient in both cysteine and methionine. However, animals are capable of converting methionine to cysteine but not vice versa.

Consequently, there is interest in increasing the methionine content of sulfurdeficient legume seeds. A range of sulfur-rich proteins has been characterized from plant seeds including the β-zeins (4.4 mol % Cys, 11.4 mol % Met), γ-zeins (over 7 mol % Cys), and δ-zeins (23-27 mol

% Met) of maize. In addition, Chui and Falco have described the molecular cloning of a protein that appears to be related to S-zeins but contains over 40 mol % methionine.

Similarly, the M, 10,000 prolamins of rice contain about 10 mol % Cys and 20 mol% Met. The methionine-rich δ-zein gene has been used to increase the methionine content of maize seeds by up to 30%, but neither this nor any other sulfur-rich cereal prolamin genes have so far been used in transgenic legumes.

Instead, the main interest in improving the sulfur amino acid content of legume seeds has been focused on the 2S albumins. Youle and Huang demonstrated that the 2S albumin fraction from Brazil nut *(Bertholletia excelsa)* contained over 17 mol % Met and 13 mol % Cys, and subsequent studies showed the presence of at least six closely related components. Methionine-rich 2S albumins have since been reported in sunflower, cotton, and amaranthus and cysteine-rich components in quinoa and pea.

Expression of the Brazil nut albumins has resulted in significant increases in the methionine content of seeds of oilseed rape, tobacco (by up to 30%), *Arabidopsis* (by 20%), narbon bean (*Vicia narbonensis*) (by up to threefold), and soybean (R. Jung, personal communication). However, it is now known that the Brazil nut protein is allergenic to humans and the commercial development of seeds expressing this protein has therefore been suspended.

Although a number of other (i.e., methionine-poor) 2S albumins are also allergenic (e.g., from mustards, castor bean, cotton), there is no evidence that this applies to the methionine-rich albumin (SFA8) of sunflower, which contains 16 Met out of 103 total residues (but see note added in proof).

Molvig et al. reported that the expression of this protein in seeds of lupin resulted in an increase in methionine by 94%, although there was no effect on total seed sulfur, the increase being at the expense of cysteine (reduced by 12%) and sulfate.

Feeding trials with rats showed that the nutritional quality of the seeds was improved significantly, with increases in live weight gain, true protein digestibility, biological value, and net protein utilization. These results demonstrate that the methionine content of legume seeds can be increased by genetic engineering but that the extent of this may ultimately be limited by the availability of sulfur within the seed.

Further increases may require the manipulation of the sulfur economy of the plant to increase the transport of sulfur into the developing grain. This will require further basic studies to identify the underlying mechanisms and their regulation. Other approaches can also be proposed to increase the methionine content of legume seeds, some of which have been tested in seeds of model species or other crop species.

Saalbach et al. showed that a methionine-enriched 7S globulin from *Vicia* was stably accumulated at levels of 1.5 to 2.2% of the total protein in seeds of transgenic tobacco plants, whereas methionine-enriched 11S globulin from the same species did not accumulate and was subsequently shown to be degraded.

Similarly, Utsumi and co-workers have shown that wild-type and methionineenriched 11S globulins of soybean accumulate stably in seeds of transgenic rice. De Clercq et al. constructed modified 2S albumins from *Arabidopsis* and chimeras comprising parts of the Brazil nut and *Arabidopsis* albumins and expressed them in seeds of tobacco, *Arabidopsis,* and oilseed rape.

Down-regulation of the sulfur-poor 1IS globulins (cruciferins) in oilseed rape has also been shown to result in compensatory increases in more sulfur-rich proteins. Similarly, Kjemtrup et al. demonstrated that mutant phytohemagglutinins of bean containing four additional methionine residues were stably accumulated in seeds of transgenic tobacco although the level was not reported. However, none of these approaches has yet been successfully applied to grain legumes.

11.8 SEED PROTEINS AND FUNCTIONALITY

Although some seeds are consumed by humans with little or no processing (e.g., boiled rice), the vast majority are processed by the food industry into a wide range of products. The ability to process seeds is determined by their functional properties, which in turn depend, to a large extent, on the seed proteins.

For example, Kinsella has identified 10 functional properties that are conferred by soybean proteins used in food systems: solubility, water absorption and binding, viscosity, gelation, cohesion-adhesion, elasticity, emulsification, fat absorption, flavor binding, and foaming.

Similarly, the viscoelastic and cohesive properties conferred by wheat gluten proteins allow flour to be processed into bread, pasta and noodles, other baked goods, and various other food products. Because these are biophysical properties determined by the structures and/or interactions of seed components, they are difficult to define in molecular terms and to manipulate.

Nevertheless, good progress has been made with the two most

important seed protein systems exploited in food processing, the gluten proteins of wheat and soybean globulins.

11.9 GEL FORMATION AND EMULSIFICATION PROPERTIES OF SOYABEAN GLOBULINS

Gelation forms the basis for making tofu, which is an important food product in the Far East. Gels result from the formation of an ordered protein network that is capable of holding water.

In contrast, emulsification properties are important for stabilizing oil-water mixtures in products such as mayonnaise, dressings, and spreads, and protein emulsifiers usually have amphipathic properties that allow them to orientate at the oil-water interface.

Utsumi and co-workers have explored the molecular basis for the gelation and emulsification properties of soybean 12S globulins (glycinins), using protein engineering and expression in *E. coli.* A range of mutations and additions, and deletions of single residues and short sequences were made, focusing on five variable regions identified on the basis of sequence comparisons.

Several mutant proteins were purified and their gelation and emulsification properties determined. These contained deletions of 11 or 8 residues from the protein N- and C-termini, respectively; the addition of blocks of four methionines into two of the variable regions; and the mutation of two cysteine residues involved in disulfide bond formation.

All mutations showed greater emulifying activity, which was increased twofold in two of the mutants, while four of them formed harder gels. Furthermore, the presence of additional methionine residues in two of the mutants raised the possibility of simultaneously improving the functional and nutritional properties.

The wild-type glycinins and one of the methionine-containing mutants have been expressed in transgenic tobacco at levels up to 4% of the total protein. The proteins were correctly processed, assembled into hexamers, and accumulated in protein bodies, although about half appeared to be partially degraded. More recent work from the same group has shown that the normal and methionine-containing glycinins can be expressed in transgenic rice seeds at levels up to about 5% of the total protein.

The protein was colocated in protein bodies with the structurally related "glutelin" storage proteins of rice and assembled into 7S (trimeric) and 11S (hexameric) forms similar to those observed in developing soybean seeds. Some of the glycinin also appeared to form hybrid oligomers with the rice glutenin. Effects on the composition and digestibility of the grain were reported but not on the functional properties.

11.10 THE VISCOELASTICITY OF WHEAT GLUTEN

The gluten proteins of wheat correspond to the prolamin storage proteins and account for about half of the total nitrogen in the mature grain. They are initially deposited in protein bodies in the cells of the developing starchy endosperm, but these deposits coalesce during the later stages of grain maturation to give, a continuous matrix surrounding the starch granules.

When white flour (i.e., starchy endosperm cells) is mixed with water to form dough, the gluten proteins form a continuous network that confers cohesive and viscoelastic properties. This network is crucial for processing; in particular, it is expanded by fermentation of leavened bread to give a light porous structure that is then "fixed" on baking.

Highly elastic (strong) wheats are required for making both bread and pasta, and poor processing quality often results from low elasticity. Wheat gluten is a complex mixture of proteins that are classified into two broad groups. The glutenins consist of individual subunits that form high-M_r (above 1×10^6) polymers stabilized by interchain disulfide bonds.

They are the major determinants of gluten elasticity, appearing to form an elastic network. In contrast, the gliadins are monomeric proteins that interact with each other and with the glutenin polymers by strong noncovalent interactions (notably hydrogen bonds). They appear to be the major determinants of gluten viscosity, acting as plasticizers. Evidence from genetic and biochemical studies indicates that one group of glutenin proteins is particularly important in determining the elasticity (i.e., strength) of gluten.

These high-molecular-weight (HMW) subunits of glutenin comprise between three and five individual proteins in hexaploid bread wheat and usually one or two proteins in tetraploid pasta wheat. Each expressed subunit accounts for about 2% of the total protein and quantitative differences in the total amount of subunit protein associated with variation in the number of expressed genes (three, four, or five in bread wheat) appear to contribute to differences in dough strength between cultivars.

However, differences in dough strength are also associated with allelic variation in expressed proteins (particularly those encoded by genes on the D genome of bread wheat). These are presumed to result from subtle differences in subunit structure and interactions. The demonstration of a relationship between subunit gene expression, protein amount, and dough strength has provided a rationale for attempts to improve gluten elasticity by introducing additional genes for HMW subunits.

This has, so far, been achieved for bread wheat in four laboratories, which have introduced genes for wild-type HMW subunits encoded by chromosomes 1 A and 1 D of bread wheat or a chimeric gene produced by combining parts of two subunits encoded by chromosome 1D.

Similarly, He et al. transformed pasta wheat with the same genes from chromosomes 1 A and 1 D of bread wheat. In all cases, the transgenes were controlled by their own promoters and expression levels up to or above those of the endogenous genes were observed.

Expression of HMW subunit transgenes resulted in increased dough strength in a poor quality bread wheat line containing only two subunits and also in pasta wheat when the transgene levels were similar to those of the endogenous genes. However, very high level expression of a chromosome 1D encoded subunit in a bread wheat line expressing five subunits or in pasta wheat resulted in unusual mixing characteristics that were similar in some respects to those of "overstrong" varieties.

In both cases the flour could be blended to increase the strength of a poor quality bread wheat flour, indicating that such lines may be of commercial value for fortification of weak flours. This work demonstrates that transformation with HMW subunits can be used to manipulate the functional properties of wheat flour, either to improve its quality for traditional end uses (bread, pasta, noodles) or to confer new properties for food (e.g., ingredients) or nonfood (e.g., packaging, films) applications.

However, a deeper understanding of the relationships between the amounts, structures, and functional properties of proteins is required to allow changes to be specifically targeted. Furthermore, attention should be paid to other groups of glutenin subunits and gliadins in addition to the HMW subunits.

11.11 CONCLUSIONS

In the present chapter we have focused on four aspects of seed composition that determine the quality for major end uses: the nutritional quality of cereals (lysine content) and legumes (methionine content) and the processing properties of wheat gluten proteins and soybean glycinin.

These examples illustrate the range of approaches that can be taken to improve quality, including the expression of genes encoding novel proteins from other plant species, the up-regulation (by inserting additional gene copies) or downregulation (by antisense or cosuppression) of endogenous genes, and the construction and expression of mutant or ab initio designed genes.

It is not surprising that amino acid composition has been one of the early successes as this can be readily defined at the molecular level. Similarly, wheat gluten and legume globulin proteins are probably the best understood seed proteins in terms of the molecular basis for their functional properties in food systems.

There are, of course, many other protein quality targets that can be addressed in the future, and there is no doubt that diet and health will have a major impact on the selection of these. Examples are the removal of proteins that are toxic, allergenic, or involved in intolerance (e.g., coeliac disease) or the introduction of novel proteins with beneficial properties. Such manipulations will be of direct benefit to the consumer and may consequently help to increase the acceptability of transgenic crops and food.

12

Genetic Engineering for Improved Nutritional Quality

Considering recent controversies about genetically modified crops and genetically modified food, one frequently asked question concerns the real need for such modifications, especially if we refer to the first commercialized traits such as herbicide or insect resistance and the possible adverse effects of their utilization on the environment.

Creating crops by transgenesis with an improved nutritional value may represent a more positive example in the public perception and would result in better public acceptance of biotechnology.

It can also be emphasized that such an approach holds a great deal of promise in terms of attaining more appropriate levels of quality food components, which can be particularly crucial for the developing world. Focusing on food nutritional quality, a series of targets can be identified.

A major goal of plant genetic engineering consists of improving the content of essential amino acids of edible crop organs to provide better food for humans and a balanced composition of feeds for domestic animals. Most of this chapter will be devoted to the progress that was been made to improve the amino acid composition of plants and in particular the lysine content.

Important results have also been reported on the qualitative aspects of a human diet, in particular vitamins, micronutrients, and phytochemicals relevant to human health. In time, plant biotechnology methods could lead to the production of feeds in which no addition of vitamins or minerals would be required.

To prevent deficiencies, the daily intake of such nutrients has to reach minimal levels that have been defined on the basis of the work of nutritionists. Unfortunately, a more or less acute lack of many of these nutrients exists in various countries of the developing world.

The diet of human populations is often essentially composed of a major staple food that does not fulfill the nutritional norms and may lead to severe illness and impair growth and development, especially in children. We intend to underline strategies that have been developed to improve the nutritive value of food by increasing the level of one or another of these nutrients and amino acids.

As such products are built up by the metabolic machinery of the plant cell, we will briefly mention a series of approaches that can be applied to redirect the metabolic flux toward increasing the level of a desired product. Then we will envisage how these methods can be applied to modify the amino acid composition of plant organs through metabolic engineering.

We will focus on the main limiting amino acids in terms of nutritional impact, essentially lysine and threonine in cereals and sulfur containing amino acids in legumes. Finally, we will describe methods of improvement beneficial to human nutrition and health with examples concerning increases of vitamin A and E levels and accumulation of deficient micronutrients such as iron.

12.1 STRATEGIES TO MANIPULATE PLANT METABOLIC PATHWAYS

The definition of possible targets for engineering a primary metabolic pathway implies a sufficient knowledge of the whole pathway in terms of genes and enzymes involved and in particular identifying the critical regulatory points in the metabolic route. In the past, such studies have been developed for a series of amino acids as exemplified by excellent reviews in books and journals.

This acquired knowledge of the factors controlling the carbon flux through the pathway, such as identification of rate-limiting steps, enzyme regulation by feedback inhibition or repression versus induction, and the presence of competitive branches in the pathway, makes it possible to establish adequate strategies for manipulating the expression level of key enzymes in the pathway with the goal of provoking changes in flux toward the synthesis of a product.

We can now modify the expression of a gene controlling a ratelimiting enzyme by insertion in the plant genome of the corresponding homologous or heterologous gene under the control of a strong promoter, expecting that a higher level of the encoded enzyme activity will be obtained.

The transgene can also be modified in such a way that it now encodes an enzyme with a higher affinity for the substrate used in the branch leading to the desired compound. A mutated gene coding for a feedback-insensitive form of the enzyme could be obtained and introduced by transgenesis. This has already been achieved for both plant and microbial genes.

Competitive branches of the pathway can be, at least partially, blocked by means of an antisense or knockout approach. Silencing the genes that exert their effects at or beyond branch nodes is thus an effective way to increase the flux toward the production of the desired metabolite by avoiding or at least decreasing a wasteful flow of intermediary products in another pathway.

The lack of success in accumulating a metabolite can be also related to the fact that its catabolic rate is increased. Here also, antisense and down-regulation approaches may allow us to decrease the effects of such catabolic degradation. These methods that rely on engineering of at least some of the genes of the pathway of interest have now been extended to the regulatory network controlling the overall pathway.

The goal is to identify master control points that can influence the expression of a series of structural genes involved in metabolic conversions in the pathway to take advantage of this global effect on the synthesis of the product of interest. Modifying the expression of a single transcription factor may affect the expression level of a battery of biosynthetic genes involved in individual branches of a complex biosynthetic pathway.

Some examples showing the potential of this approach are already available in the case of secondary metabolism (69%) and also amino acid biosynthesis. The increasing availability of regulatory genes and the dissection of signal transduction pathways influencing primary metabolism make this type of methodology particularly attractive for the biotechnologist. Finally, we also have to take into account the complexity of the biosynthesis we wish to orientate toward a defined direction.

In that context, the development of functional genomics will provide us with invaluable information about the changes at the RNA, protein, and metabolite levels that occur when we modify defined steps of a specific pathway. It will also allow us to monitor the introduced changes so that desired alterations remain compatible with the growth and development of the crop.

Moreover, we have to mention the possibility of engineering not only nuclear genes but also organelle genomes, in particular in the case of plant chloroplastie DNA. Here we can take advantage of the impressive increase in the expression of an inserted gene due to the large number of copies present in each leaf cell of the plastid transformed plant.

As many steps of amino acid biosynthesis have been located inside the chloroplast, this represents a new avenue full of promise for overexpression of amino acid metabolism. All the forecast benefits of metabolic engineering may still be limited or annulled by the complexity of the regulation controlling many plant biosynthetic pathways that still have to be worked out before we can obtain the production of the desired metabolites systematically and reliably.

Besides manipulating gene expression of a pathway, the compartmentation of the metabolites in different cellular compartments and their transport inside the cell and to the different plant organs can totally alter the results of manipulating the properties of one key enzyme in the pathway.

In particular, better knowledge of how quantifying and directing metabolic flux obtained by nuclear magnetic resonance spectroscopy appears essential.

12.2 MANIPULATING ESSENTIAL AMINO METABOLIC PATHWAYS

The nutritional quality of crop plants is determined by their content of essential amino acids provided as proteins in foods for humans or in feed for monogastric animals. Plant proteins are often deficient in some of the 10 essential amino acids that are required in the human and animal diet.

In general, cereals are deficient in lysine and legumes in the sulfur amino acids, methionine and cysteine. Several approaches have been presented and used with the goal of increasing the relative content of essential amino acids in plants, in particular the aspartate-derived amino acids, lysine and threonine. One of the earliest approaches consists of screening genotypes and induced mutants to identify plants with higher lysine levels in the seed.

This method was first applied successfully to maize and led to the high-lysine opaque-2 type (Later, similar types of mutants were reported in barley and *Sorghum.* The increase in total lysine content was due to a shift in the spectrum of storage proteins to the benefit of lysinerich albumins and globulins instead of lysine-poor prolamines.

However, this type of mutation was accompanied by a drastic reduction in yield and protein content. The introduction of gene transfer technologies opened new ways to alter the amino acid composition of the seed proteins, in cereals and legumes in particular. In this case, genes encoding storage proteins naturally rich in essential amino acids or engineered genes leading to the production of "synthetic" proteins displaying a high content of defined essential amino acids were expressed in transgenic plants under the control of suitable promoters.

In order to modify the nutritional value significantly, the foreign proteins must constitute a high percentage of the total proteins, so that in addition to high levels of transcription and translation, the stability of the additional proteins is crucial.

As underlined by Galili et al., although numerous attempts founded on such strategies have been reported, a significant, reliable accumulation of a stable nutritionally valuable protein still has to be convincingly demonstrated. In that context, it is worthwhile to mention the work of Chakraborty et al. on the expression in potato of an *AmA-1* gene coding for a seed-specific albumin from *Amaranthus hypochondriacus* with a well-balanced amino acid composition.

The authors reported that when this gene was put under the control of a granule-bound starch synthase promoter, a striking increase in most essential amino acids present in potato tubers was observed. Unexpectedly, the growth and production of tubers as well as the total protein content of this accumulation organ were significantly increased.

An alternative method consists of producing plants containing elevated levels of. specific amino acids by manipulating their metabolism. This can be achieved through alteration of key enzymes from those biosynthetic pathways by mutation and transformation.

We have focused on the aspartate derived amino acid biosynthetic pathway producing the four essential amino acids lysine, threonine, isoleucine, and methionine. Considerable research, aided by genetic engineering, has been devoted to this pathway, its enzymes, and their regulation with the aim of optimizing amino acid content.

12.2.1 Biochemical Regulation of Lysine and Threonine Biosynthesis

Higher plants synthesize lysine, threonine, isoleucine, and methionine

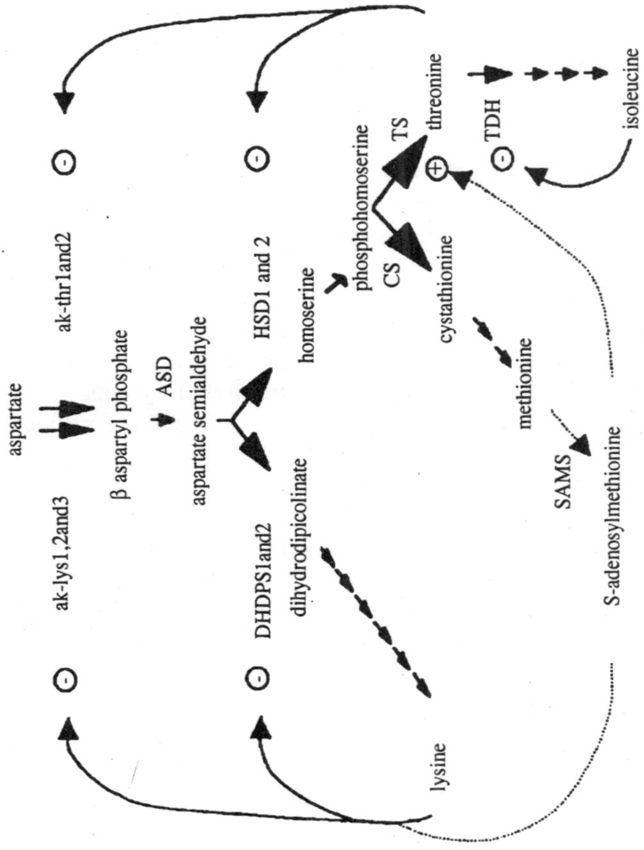

Figure 12.1: The aspartate family biosynthetic pathway.

from the common precursor aspartate via a complex branched pathway. Understanding the inner regulation taking place during this process is essential with respect to knowledge of basic plant metabolism but also to possibly modifying the nutritional value of crops. Close examination of the lysine and threonine biosynthesis revealed the existence of several enzymatic steps feedback controlled by these two end products.

Essentially, two enzymes play key regulatory roles in the pathway: aspartate kinase (AK), the first enzyme of the overall pathway, which is retroinhibited by either lysine or threonine, and dihydrodipicolinate synthase (DHDPS), the first enzyme of the lysine-specific branch, which is severely inhibited by lysine.

The main limiting step in lysine production appears to occur at the DHDPS level via the feedback control exerted by lysine (K_i, between

5 and 20 μM). Most of the biosynthetic enzymes of the pathway have been localized in the plastids on the basis of biochemical data an i molecular evidence from genes encoding the majority of the biosynthetic enzymes, in which putative chloroplastic transit peptides have been identified.

12.2.2 Mutants Overproducing Free Lysine or Threonine

We will now focus on an approach based, as previously mentioned, on the deregulation of biosynthetic pathways of such essential amino acids, in particular in the case of lysine and threonine. In various species, mutants accumulating lysine or threonine in the soluble amino acid pool have been isolated by selecting for growth in the presence of the lysine amino acid analogue S-2-(aminoethyl)-1-cysteine (AEC) or toxic combinations of lysine and threonine in the culture medium.

Although threonine overproduction was detected in all tissues analyzed and in particular in seeds, the increase of free lysine was noticeable only in leaves and calli. Moreover, an aberrant phenotype characterized by reduction of the foliar surface, absence of shoot elongation, and sterility was observed when high levels of free lysine accumulated (25% and more of the total free amino acid pool).

The accumulation of both amino acids in leaves is also dependent on the developmental stage, reaching a maximum at shoot elongation. The biochemical characterization of these mutants has shown that we are dealing with deregulation at the level of the two key enzymes, DHDPS in the lysine overproducer mutant and AK in the threonine overproducer. These mutated forms are both fully insensitive to feedback inhibition by the respective amino acid.

12.2.3 Key Genes of the Lysine and Threonine Pathways

12.2.3.1 Aspartate Kinase Genes

At least five forms of aspartokinase have been identified in plants and may be.classified on the basis of their differential feedback properties. One gene coding for a threonine-sensitive form (ak_1-hsd,) was isolated by heterologous hybridization *in Arabidopsis thaliana.* The deduced amino acid sequence revealed the presence of a second region corresponding to the enzyme activity of homoserine dehydrogenase (HSD), the first committed enzyme in the branch of the pathway that leads to threonine synthesis.

We are thus dealing with a gene coding for a bifunctional protein with an AK activity at the NH_2-side and an HSD activity at the COOH-terminus. The expression of the *A. thaliana* ak_1-hsd_1 gus gene was elevated in an array of young plant tissues containing actively growing cells (meristems, young leaves, cortical and vascular stem tissues, anthers, gynoecium, developing seeds).

During the development of the embryo its expression appeared coordinated with the initiation and onset of storage protein synthesis. A second gene showing a high identity (82%) with the ak_1-hsd_1 nucleotide sequence was identified in a bacterial artificial chromosome (BAC) clone and localized as ak_2-hdh_2 on chromosome 4 of the *A. thaliana* genome. As the major part of the AK activity is sensitive to feedback inhibition by lysine, attempts have been made to clone the corresponding genes.

Degenerated primers corresponding to conserved motifs between lysine-sensitive bacterial AKs were used to clone two genes *(ak-lys1* and *ak-lys2)* encoding monofunctional AKs in *A. thaliana.* The presence of two nuclear genes both encoding AK-lys proteins and both targeted to the chloroplast, as shown by the presence of a transit peptide, but with only 70% identity at the deduced amino acid level raises questions about their respective roles in plant development.

Therefore, A. *thaliana* and *Nicotiana tabacum* have been transformed with the 5' upstream region of either *ak*lys gene fused with the *uidA* reporter gene encoding glucuronidase (GUS). In *A. thaliana*, a comparison of the GUS patterns of *ak-lys1* and *ak-lys2* revealed strongly predominant expression of *ak-lys2* over *ak-lys1*.

Throughout the vegetative phases, *ak-lys2-GUS* expression reached a high level, especially in the vasculature, whereas *ak-lys1-GUS* plants showed a markedly reduced intensity of the histochemical staining. In the reproductive phase, both genes were well expressed in flowers, *ak-lys2* was the only one expressed in the fruits, and no staining could be detected in seeds in both cases.

The physical position of *ak-lysl* and *ak-lys2* could be assigned to chromosome 5 at two different loci at a distance of 2 cM. The *Arabidopsis ak-lys* gene family is thus composed of at least two different members. We identified a third *ak-lys* gene located on chromosome 3 of *Arabidopsis* using the Blast program to screen databases. The deduced *ak-lys3* amino acid sequence showed higher identity with the *ak-lys2* (82.8%) than with the *ak-lysl* (*68%*) peptide sequence. The *ak-lys3* gene was reported to be highly expressed in leaf vein tissues.

The isolation and characterization of *A. thaliana* mutants displaying an AK-HSD isozyme less sensitive to threonine inhibition or an AK-lys isozyme less sensitive to lysine feedback inhibition will make it possible to identify at the nucleotide level mutations that lead to threonine accumulation.

After incorporation into the corresponding copy DNA (cDNA), these mutated plant genes can be respectively expressed in an appropriate construct to obtain threonine accumulation in specific tissues, especially in seeds.

12.2.3.2 Cloning and Characterization of Wild-Type and Mutant Genes

A mutant of *Nicotiana sylvestris* (RAEC-1) was shown to overproduce lysine because of a mutation in the DHDPS gene that causes the DHDPS enzyme to be insensitive to the normal feedback inhibition of lysine. The dhdps-rl mutation was identified as a substitution of two nucleotides changing asparagine in isoleucine in a conserved region of the protein.

In maize, a series of single amino acid substitutions were found to eliminate lysine inhibition of DHDPS. DHDPS-encoding sequences were cloned by functional complementation in a bacterial DHDPS-deficient strain. This first isolated clone made it possible to obtain and sequence a full-length *Arabidopsis* DHDPS cDNA.

Constructs derived from the *Arabidopsis* cDNA allowed the generation through ethyl methane sulfonate in vitro mutagenesis of clones encoding fully insensitive forms of the DHDPS protein. Furthermore, the clones successfully isolated by functional complementation in a DHDPS-deficient *E. coli* mutant were all found to encode insensitive enzyme forms, which means that complementation selects for insensitive DHDPS plant enzymes.

In soybean, three mutants were constructed containing specific amino acid substitutions that lead to lysine-desensitized DHDPS. In a further step, the *Arabidopsis* promoter has been isolated and fused with the reporter gene gus to study the transcription properties of this upstream sequence.

Expression of GUS was detected in meristems and vascular tissues of roots, in vascular tissues of stems and leaves, and in the meristems of young shoots. In flowers, high expression was found in the carpels, pollen grain, and young embryos but not in endosperm of mature seeds. No lysineinduced repression of the *dhdps* gene could be detected. An *Arabidopsis* genomic sequence encoding a second DHDPS enzyme was identified by screening the EMBL database.

The *dhdps-2* coding sequence shows 84% identity with the nucleotide sequence of *dhdps-1*. The genomic *dhdps-2* sequence contains three exons and two introns, whereas only one intron and two exons are present in *dhdps-1*. Comparison of the promoter regions of *dhdps-1* and *dhdps-2* did not reveal boxes with any significant conservation.

The two *Arabidopsis* genes were localized on chromosome 2 for *dhdps-2* and chromosome 3 for *dhdps-1*. That *dhdps-2* encodes a functional protein was shown by the growth on a nonsupplemented medium of an *E. coli* DHDPS-deficient strain transformed with the *dhdps-2* apoprotein coding sequence.

Activity tests performed in the presence of increasing concentrations of lysine proved that the DHDPS-2 enzyme is also strongly feedback inhibited, with a 50% loss of activity at 30 μM lysine. Via promoter-GUS fusion, expression of the *dhdps-2* gene was observed during the whole developmental cycle and appears to be quite similar to that observed for the *dhdps-1* gene, although the *dhdps-2* gene is in general a little more expressed. DHDPS-2 activity was strongly detected in the vasculature of stems and leaves, in carpels, and in developing seeds.

12.2.4 Improving Lysine and Threonine Accumulation via Transfer of Bacterial Genes

The availability of bacterial and plant genes encoding feedback-insensitive enzymes allowed redirecting the expression of these genes in plants and in particular in storage organs. Initial experiments made use of constructs monitored by the strong cauliflower mosaic virus (CAMV) promoter harboring a bacterial *dapA* gene that encodes a DHDPS partially or fully feedback insensitive and mutated alleles of the *E. coli lysC* gene encoding a feedbackinsensitive AK.

Transgenic lines have been obtained in tobacco, potato, barley, *Arabidopsis,* and oilseed rape and soybean. As previously mentioned for *Nicotiana sylvestris* mutants, expression in tobacco of the bacterial AK led to accumulation off free threonine not only in vegetative tissues and in tubers in the case of potato but also in reproductive organs and seeds.

In the case of the bacterial DHDPS, the lysine overproduction, although present in vegetative tissues, was not expressed in mature seeds. Constitutive overexpression of the bacterial DHDPS gene was also accompanied by phenotypical alteration of the transgenic plants, as observed with the selected *N. sylvestris* mutants, at least when a high lysine level was reached.

Targeting the expression of these bacterial genes to sink organs such as seeds was then proposed as a way to alleviate such deleterious effects. Expression of the chimeric AK and DHDPS gene under the control of the β-phaseolin promoter obtained from the bean gene coding for this seed protein was thus evaluated.

Transgenic tobacco seeds showed an important increase in free threonine but no significant change in the amount of lysine in mature seeds, although the activity of bacterial feedbackinsensitive DHDPS was clearly higher than the DHDPS activity measured in nontransformed plants.

This lack of lysine accumulation in tobacco seeds was then ascribed to enhanced lysine catabolism as shown by higher activities of the major lysine catabolism enzyme lysine-ketoglutarate reductase (LKR). We have to underline that the lysine content present in mature seeds of transgenic plants appears to be the result both of increased accumulation due to high activity of a feedback-insensitive DHDPS and of the rate of lysine catabolism.

According to the balance between enhanced biosynthesis and induced catabolic degradation, the accumulation of lysine in seeds may vary among plant species. This is best shown by the results obtained by Falco et al. and Mazur et al. when similar bacterial ak and *dhdps* genes were transformed into oilseed rape and soybean. Alone or combined with AK, expression of the bacterial DHDPS resulted in a dramatic increase of free lysine (10-100 times the level of nontransformed plants) with a significant effect on the total lysine content, which increased more than twice.

In these last cases, accumulation of catabolic products such as saccharopine and α-aminoadipic acid was eventually observed but only at a minor level. However, in maize, when the bacterial AK and DHDPS were expressed with an endosperm or an embryo-specific promoter, lysine accumulation was detected only in the embryo but was sufficient to raise the overall lysine concentration in seed by 50 to 100%.

Bacterial enzymes can thus be expressed with success in vegetative and sink organs such as seeds leading to deep modifications in lysine metabolism.

12.2.5 Improving Lysine Content Using Plant Gene Transfer

The cloning of mutated plant genes provided similar tools for metabolic engineering of plants. With the goal of increasing lysine production, several chimeric constructs harboring the *dhdps-r1* mutated allele have been developed under the control of the 35S CAMV promoter as well as tuber- or seed-specific promoters and transferred in a model species *(Nicotiana plumbaginifolia)* and crops such as *Solanum tuberosum* and *Sorghum bicolour.*

Table 12.1: Levels of Lysine Overproduction in Seeds of Some Transgenic Crops

		Lysine	
Plant species	***Inserted genes***	***Free as***	***Total as***
Brassica napus	Corynebacterium dhdps	3.1-140	1-1.75
Brassica napus	Colynebacterium dhdps + *E. coli ak*	4.7-38	1-2.0
Glycine max	Corynebacterium dhdps	12-25	1.25
Glycine max	Coiynebacterium dhdps + *E. coli ak*	335	0.9-4.7
Hordeum sativum	*E. coli dhdps*	2	1.05

In *N. plumbaginifolia,* overexpression of the mutated *dhdps-rl* gene led to significant lysine overproduction in vegetative tissues, flowers, and immature seeds. Thus, the insensitivity of the mutated enzyme to feedback inhibition is critical for accumulating lysine, but the level of the enzyme also plays an important role in the increased lysine synthesis. Transformants with high DHDPS activities are also characterized by high free lysine levels.

The K_m and V_{max} of DHDPS thus represent determinant factors in the monitoring of lysine biosynthesis. Constitutive overproduction of this insensitive DHDPS was accompanied by abnormal phenotypes (multiple shoots, linear leaves, partial sterility) at least when high free lysine contents were reached in the transgenic plants, as already mentioned for lysine overproducer mutants of *Nicotiana sylvestris* and transgenic tobacco plants. In potato, the same gene was introduced under the control of a patatin promoter to confine the lysine overproduction to tubers.

In 23 analyzed tranformants, the range of lysine content varied from 0.9 to 13.7% of the total free amino acids while the control variety displayed a mean value of 2.6%. Thus, transformation of potato resulted in a fivefold increase in free lysine content without any noticeable modifications in tuber structure and, production. The properties of transgenic sorghum plants expressing the mutated *dhdps-rl* gene of *N. sylvestris* under the control of the 35S CAMV promoter were also determined.

Taansgenic plants were produced by introduction of the gene into

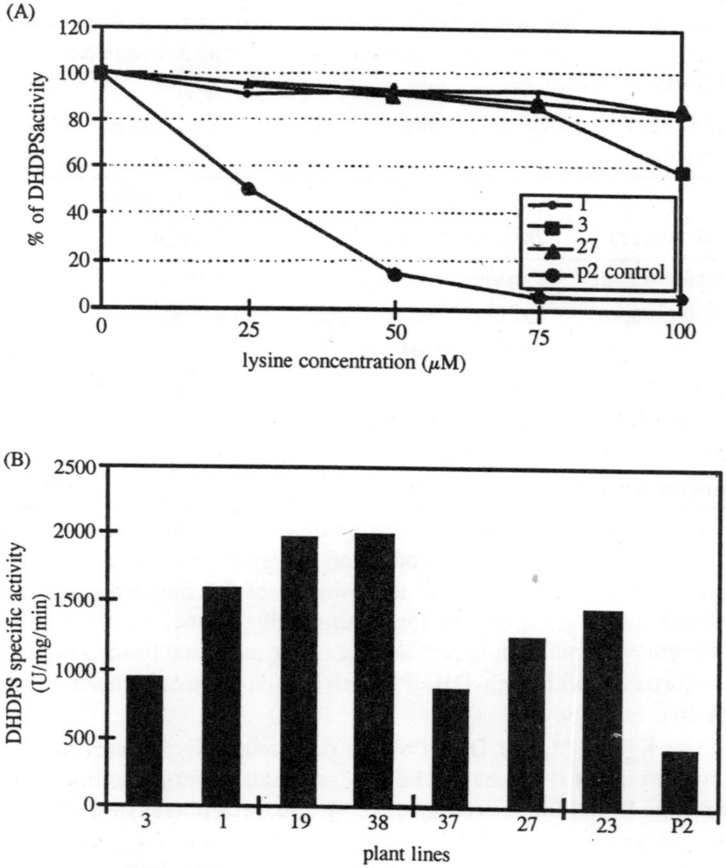

Figure 12.2: (A) Effect of lysine on DHDPS activity in leaf extracts of transgenic and wild-type (P2) Nicotiana plumbaginifolia plants. (B) DHDPS activity in leaves from transgenic lines and wild-type P2 plants.

Sorghum immature embryos by microprojectile bombardment. These embryos were then selected on the lysine analogue 2-aminoethyl-Lcysteine (AEC) from callus initiation up to regenerated roots.

As a result of the expression of the gene, an active DHDPS enzyme insensitive to feedback inhibition was produced in the primary transformants and their progenies. In leaves of transgenic plants, a low but significant increase of lysine (1.5 to 2.5 times more than the control) could be associated with the expression level of the ectopic *dhdps-r1* gene.

Sorghum seeds also synthesized higher levels of free lysine than observed in the original cultivar during the first phase of their development, from 15 to 25 days after pollination. However, toward maturity the lysine content of transgenic seeds decreased and was almost comparable to the control value.

This evaluation during maturation can be ascribed to enhanced lysine catabolism, as already mentioned for transgenic tobacco plants.

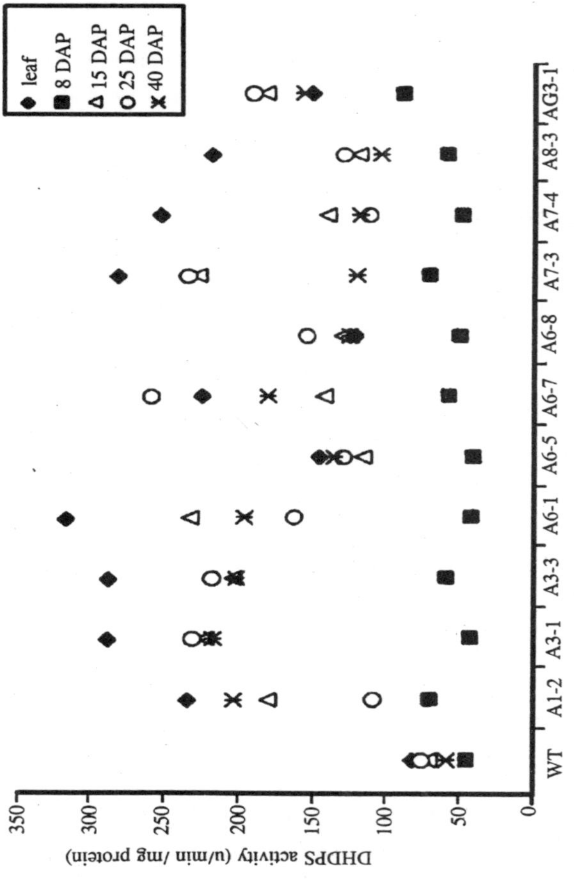

	WT	A1-2	A3-1	A3-3	A6-1	A6-5	A6-7	A6-8	A7-3	A7-4	A8-3	AG3-1
Flag Leaves	1.8	4.3	3.8	3.5	3.2	3.4	3.7	3.1	3.8	3.9	3.5	4.3
8 DAP	1.7	1.9	2.1	1.5	1.4	1.6	1.2	1.4	1.6	2.0	1.2	1.4
15 DAP	1.8	2.9	3.9	3.9	7.1	4.9	5.6	2.5	3.8	2.1	2.4	2.1
25 DAP	1.6	3.0	3.1	4.6	4.1	2.9	2.6	4.6	2.5	3.2	1.9	2.7
40 DAP	1.9	2.5	1.4	2.2	2.0	1.1	1.8	2.4	2.1	ND	1.6	ND

percentage of free lysine content

Figure 12.3: DHDPS specific activity measured in flag leaves and seeds of *Sorghum,* harvested on wild-type plants (WT) and T2 progenies of transformants (from A 1-2 to A G-3) at different developmental stages. Seeds were collected at 8, 15, 25, and 40 days after pollination (DAP) and were analyzed for DHDPS activity. Values at the bottom represent the percentage of lysine in the free amino acid pool. Each value is the mean of three independent assays from three individual T2 plants in each line.

A similar observation has been reported in corn, in which the expression of the deregulated bacterial *dhdps* gene under the control of an endosperm-specific promoter did not lead to any lysine accumulation.

The presence of LKR activity has been demonstrated in immature endosperm in maize and also in *Sorghum* seeds in our laboratory. In conclusion, these results show clearly that although we are dealing with a complex, highly regulated biosynthetic pathway leading from aspartate to four essential amino acids, the overexpression of a single gene encoding a feedback-insensitive form of the key DHDPS enzyme exerts a significant effect on the carbon flux through the aspartate pathway toward lysine.

This accumulation of lysine was in some species accompanied by phenotypical alterations, which should be avoided when this method is applied to crops. Organ-specific overproduction of lysine and threonine through the use of specific promoters can overcome the limitations due to the constitutive expression of deregulated genes in transgenic plants.

Although lysine synthesis can be increased in plants, the level of free lysine in mature seeds may be determined not only by the efficiency of the biosynthetic pathway (in terms of activity of a feedback-insensitive DHDPS) but also by the stability of the amino acid accumulated in the free pool.

In this case, the increased lysine catabolism due to the induction of LKR in endosperm tissues is a limiting factor leading to the reduction of free lysine accumulation. The genes encoding the two first catabolic enzymes, LKR and saccharopine dehydrogenase, have been cloned from *Arabidopsis* and maize.

Therefore, by using an antisense or cosuppression approach, it might be possible to reduce the expression of LKR and the buildup of lysine degradation products. Another complementary approach will be to produce transgenic plants, in particular cereals, that altogether overproduce free lysine and express genes encoding lysine-rich proteins in seeds.

This would provide a sink for stable accumulation of the increased lysine supply as bound amino acid in the endosperm lysine-rich proteins, avoiding the induction of strong catabolic activity.

12.2.6 Manipulating Plant Micronutrients to Benefit Human Health and Nutrition

Plant foods also contribute to ensuring sufficient essential vitamins and minerals for the human diet. Such micronutrients are often poorly represented in staple crops, and attempts have been developed to increase the synthesis of these compounds by isolating genes required for their synthesis and overexpressing them under the control of pertinent promoters. Lack of vitamin A and vitamin E and iron deficiency are common problems in developing as well as in developed countries, involved in a series of illnesses, particularly in children.

In the major staple food rice, provitamin A is not synthesized in the endosperm, which represents the main part of the seed. The latest precursor of the provitamin A carotenoid is geranylgeranyl diphosphate (GGPP). Theoretically, the synthesis of β-carotene from this early intermediate would require the expression of four additional enzymes: phytoene synthase, phytoene desaturase, carotene desaturase, and lycogene β-cyclase.

A phytoene synthase gene originating from daffodil was associated in a vector with the sequence coding for a bacterial phytoene desaturase that is capable of introducing the required double bonds. Lycopene β-cyclase from daffodil was carried by another vector, and cotransformed rice plants were selected.

The relevant genes were placed under the control of an endosperm-specific glutelin promotor or the constitutive CAMV 35S promoter in the case of phytoene desaturase. Transgenic rice harboring this combination of genes produced seeds with a yellow endosperm indicating carotenoid formation.

Biochemical analysis confirmed the presence of provitamin A at a level of at least 2 μg/g in homozygous lines, which should ensure the necessary daily amount of vitamin A in a typical Asian rice diet (300 g of rice per day). The grain known as golden rice is expected to be beneficial to people suffering from vitamin A deficiency. In the case of vitamin E, the example illustrates the power of a genomics-based approach.

The final step of the pathway is of special interest here because the methylation of γ-tocopherol to α-tocopherol, the biologically active form of the vitamin E, is performed by γ-tocopherol methyl-transferase (γ-TMT), whose activity is a limiting factor in many seeds. Therefore, the objective was to clone the corresponding gene and to overexpress it in seeds.

This was made possible by a genomics-based approach by identifying probable orthologs of the γ-TMT of *Arabidopsis* using the protein sequence of a cyanobacterium *Synechocystis,* which also synthesizes α-tocopherol. In transgenic *Arabidopsis* plants overexpressing the γ-TMT enzyme, more than 95% of the total tocopherol pool was in the form of α-tocopherol whereas the wild-type seeds contain practically only γ-tocopherol.

Similar manipulations can now be used to improve the vitamin E level of important crops such as rapeseed, maize, and soybean. Iron deficiency is also a significant nutritional problem for 30% of the world population. Rice endosperm contains only low amounts of iron, which, combined with a high content of phytate (an inhibitor of iron resorption in the intestine), causes a dietary iron deficiency in populations with a rice-based diet.

One approach has been to express under the control of glutelin promoter a soybean gene coding for a ferritin protein used as a storage form of iron. Transgenic seeds have been shown to accumulate three times more iron than seeds from a control line.

With this modification can also be associated the reduction of phytate concentration in cooked rice via a transgene coding for a thermotolerant phytase from *Aspergillus fumigatus*. Both vitamin A and iron-rich traits can be combined by sexual crossing, and such material is available for noncommercial use to developing countries. The low availability of phosphorus (P) in many soils is an important factor limiting agricultural production worldwide.

In an approach for developing novel crop varieties more efficient in the use of P, Lopez-Bucio et al. engineered plants able to produce more organic acids and in particular citrate by overexpressing a bacterial citrate synthase coding sequence in tobacco.

Citrate-overproducing plants yield more leaf and fruit biomass when grown under P-limiting conditions and require less P fertilizer to achieve optimal growth. In conclusion, we think that in time genetic engineering may lead to varieties carrying all the additional vitamins and minerals required in foods and feeds.

12.3 CONCLUSIONS AND PERSPECTIVES FOR THE FUTURE

Transgenic crops with improved nutritional value have already been produced by overexpressing homologous, heterologous, and eventually mutated genes involved in key steps of the metabolism of amino acids and vitamins. In the case of lysine and threonine, overproduction of these essential amino acids has been achieved by the expression of feedback-insensitive AK and DHDPS enzymes.

We stress that the overexpression of a single gene encoding a feedback-desensitized DHDPS exerts a significant effect on the flux through the aspartate-derived amino acid pathway toward lysine. With regard to metabolic engineering, this example is quite notable although direct extrapolation to other metabolic pathways will not necessarily meet with success.

Rigidity of the metabolic network, functional redundancy of genes, and posttranscriptional control processes can alter the expected results and lead to undesirable side effects. The desired metabolite may be degraded, limiting its accumulation, as exemplified for lysine in seeds of at least some plant species. Another constraint concerns the number of genes that have to be transferred in the recipient plant genome to reach or/and optimize the accumulation of a valuable product.

Coordinated alterations of rate-limiting enzymes associated with the overexpression of master regulatory genes will probably contribute to increasing the metabolic flux toward the desired metabolite. However, the expression of multiple genes involved in a complex biosynthetic pathway will require the development of vector systems and methods ensuring the integration and stable expression of gene batteries in the plant genome.

Further progress in enhancing the nutritional quality of crops will also rely on emerging technologies, and we can anticipate decisive advances through functional genomics, DNA microarray technology, and metabolic flux analysis to obtain more insight into metabolic networks and their regulation in plants.

It is important to obtain more information on metabolic sites of synthesis and storage and on metabolite transport between organs to learn how to exploit the possibilities offered by biochemical engineering and to apply them to metabolic processes relevant to the nutritional quality of crops.

13

Genetic Engineering in Manipulating Metal Accumulation

Soils and waters with high levels of toxic metals such as cadmium, arsenic, lead, and mercury are detrimental to human and environmental health. Toxic heavy metals contaminate soils and waters in industrialized nations as well as in developing nations.

The four heavy metals arsenic, lead, mercury, and cadmium have been identified as belonging to the five priority most hazardous substances found at toxic Superfund sites in the United States. Many human disorders have been attributed to ingestion of heavy metals including learning disabilities in children, dementia, and increased rates of cancer in response to cadmium (Cd).

Removal of heavy metals from highly contaminated soils and waters is therefore a very costly but necessary process that is currently being pursued at contaminated sites worldwide. Remediation of soils containing high levels of toxic heavy metals is pursued by physical removal of metals because most of these metals cannot be degraded in the soil, in contrast to many organic contaminants.

Current practical methods used to decontaminate such sites involve physical excavation of top soils, transport, and reburial elsewhere. However, these cleanup methods are feasible only for small soil areas and are very costly. For example, costs for cleaning 1 ha to a depth of 1 m have been reported between $600,000 and $3,000,000 (5). Furthermore, excavation strategies are not applicable to contaminated waters. Alternatively, metals can be immobilized in soils.

This approach can carry the long-term risk of resolubilization because of chemical or biological changes (e.g., acidification). Complementary approaches involving both removal and immobilization will be needed to remediate heavy metal-contaminated sites. Research and applications indicate that uptake of heavy metals into plants via the root system could provide a cost-effective approach for toxic metal removal and remediation of heavy metal-laden soils and waters.

Plant roots have been shown to remediate waters by removal of heavy metals. Furthermore, increasing the accumulation of valuable metals could eventually lead to "phytomining," as proposed, for instance, for gold. On the other hand, uptake of toxic heavy metals in crop plants is known to cause health problems. Understanding of the mechanisms underlying toxic metal uptake and sequestration could lead to engineering of crops that avoid toxic metal accumulation in edible parts of plants.

Toxic heavy metals are transported across the plasma membrane into plant root cells via physiological metal uptake transporters. However, for plants to accumulate large amounts of toxic metals for bioremediation purposes, many mechanisms and genes need to be identified and modified in plants.

Several rate-limiting steps are critical for effective removal of heavy metals from soils. These include making the contaminants biologically accessible in the soil by chelation or external acidification and subsequent transport of metals or complexed metals across the plasma membrane of root cells.

Upon uptake of heavy metals into plant cells, intracellular detoxification and transport through plant tissues are required. Other metals, such as iron and zinc, are essential nutrients. Iron deficiency is the most widely spread micronutrient deficiency worldwide, affecting up to 3.7 billion people, particularly women, and zinc deficiency is a significant limiting factor for production and quality of cereals.

Identification of molecular mechanisms that enhance accumulation of essential nutrients in plants could lead to development of crops that help to address major nutritional problems in humans and could also play an important role in plant nutrition and growth. Phytoremediation involves the biotechnological exploitation of metal tolerance mechanisms.

The same applies for the nutrition-related objectives of minimizing the concentration of toxic metals in edible parts and enriching crops for essential metals. In the present chapter we summarize the molecular understanding that has been gained to date and review the transgenic approaches that have been pursued to increase plant metal tolerance.

This includes work that does not exploit plant tolerance mechanisms but instead utilizes bacterial or mammalian genes.

The heavy metals mainly considered are cadmium, mercury, copper, zinc, iron, and nickel. Also, even though aluminum, strictly speaking, is not a heavy metal (because the specific weight is under 5 g/cm^3), we include work on Al tolerance because of its agronomic importance.

Most of the fundamental insight into metal homeostasis in eukaryotes has come from studies involving. *Saccharomyces cerevisiae*. For many of the molecular components identified in yeast, plant and mammalian homologues have been found, indicating a high— degree-of conservation. Thus, the work on *S. cerevisiae* is covered in some detail.

13.1 TOLERANCE MECHANISMS

The toxic effects of plant exposure to elevated metal concentrations, such as growth inhibition and chlorosis, are well documented. However, as for other organisms, the biochemical understanding of metal toxicity is limited. Class B metals such as Cu(I) and Hg and borderline metals such as Cu(II), Zn, Ni, and Cd (12) can form complexes with nitrogen and sulfur atoms in proteins, thereby potentially inactivating them.

The nonessential metals Cd and Pb compete for binding with the essential metals Ca and Zn. Also implicated in symptoms of metal toxicity is the redox activity of some metals (e.g., Cu, Fe) that may result in the formation of reactive oxygen species. A complex network enables plants to control tightly the intracellular concentrations and distribution of essential heavy metals such as copper and to minimize the cytosolic concentrations of nonessential heavy metals such as cadmium.

The interplay of mainly transport and chelation processes that constitutes this network results in a "basic metal tolerance," the distribution, sequestration, and exclusion of potentially toxic heavy metal ions. In addition, a number of plant species show "metal hypertolerance." They can grow on soil that naturally or because of human activities contains heavy metal concentrations that are growth prohibiting to most plants.

These species belong to a specialized flora that has colonized Ni-rich serpentine soils or Zn- and Cd-polluted areas. Two principal responses to otherwise toxic heavy metal concentrations can be found, metal exclusion and metal hyperaccumulation, with the latter mainly restricted to Ni, Zn, and Se. About 400 different species belonging to a wide range of taxa have been described as hyperaccumulators, about 75% of which are Ni hyperaccumulators.

It is a general characteristic of both basic metal tolerance of most organisms and hypertolerance of some specialized plants that most of the mechanisms involved appear to be specific for a certain metal or a small group of metals. For instance, fission yeast deficient in phytochelatin synthesis are Cd and Cu hypersensitive yet grow normally, when compared with wild-type cells, in the presence of elevated Zn concentrations. Some Ni-hyperaccumulating plants show the so-called histidine response, which mediates Ni tolerance.

The sensitivity toward most other heavy metals, however, is not attenuated by the increases in histidine levels. Five general metal tolerance mechanisms can be distinguished: chelation, sequestration, exclusion, biotransformation, and repair.

The first three of these categories constitute the principal elements of the metal homeostasis network. Exclusion includes the regulation of metal uptake into cells. Biotransformation, i.e., reduction and possibly volatilization, appears to be restricted mainly to Hg and the metalloids Se and As.

13.1.1 Chelation

Because of their reactivity with N, S, or O, it has been postulated that metal ions are constantly chelated when they are taken up into the cell by specialized or unspecific cation transporter. By studying the interaction of the Zn/Cu superoxide dismutase, its copper chaperone, and other chelators, it was demonstrated for *S. cerevisiae* that the intracellular concentration of free copper is indeed limited to less than one atom per cell. Specialized chelators exist that function as cytosolic mediators of metal trafficking, the delivery of essential metal ions to organelles and protein complexes.

13.1.1.1 Phytochelatins and GSH Metabolism

The best-known chelators in plant cells are phytochelatins (PCs), small metal-binding peptides that are nontranslationally synthesized from glutathione or glutathione derivatives such as hydroxymethyl-glutathione in a transpeptidation reaction. PCs of chain lengths up to $n = 11$ are formed in plants and some fungi upon exposure to a range of different metals.

However, not all of the inducing metals are chelated by phytochelatins. Well characterized is the formation of low-molecular-weight cytosolic PCCd complexes, which are transported into the vacuole, where high-molecular-weight complexes are formed on incorporation of S^{2-}. Also, in vitro studies showed that Zn and Cu ions could be transferred efficiently from Zn- or Cu-PCs, respectively, to metal-requiring apoenzyme. The importance of PC synthesis was demonstrated by the isolation of the *Arabidopsis cad1* mutant. This mutant is PC deficient and Cd hypersensitive. The *CAD1* gene and other plant and fungal genes encoding phytochelatin synthases (PCSs) have been cloned by positional cloning and based on the screening for Cd tolerance mediating plant complementary

DNAs (cDNAs) in *S. cerevisiae*. The analysis of a *Schizosaccharomyces pombe PCS* knockout strain and of *S. cerevisiae* cells expressing PCS genes from different sources showed that the formation of PCs confers Cd tolerance and to a limited degree Cu, arsenate, and Hg tolerance. Based on these yeast studies, there is no indication for a role in, for instance, Zn, Co, or Ni detoxification. In the meantime, PC-mediated As tolerance was confirmed for plants.

The expression of PCS genes in baker's yeast, which resulted in a more than 15-fold increase in Cd tolerance, exemplified the potential of PC synthesis to enhance Cd tolerance in organisms that do not normally produce PCs.

However, the situation may be different in plants because apparently all plant species express phytochelatin synthases. There have been numerous attempts to enhance PC formation and thereby plant metal tolerance by generating plants overproducing the PC precursor glutathione. *Brassica juncea* plants overexpressing the *Escherichia coli gshl* gene, encoding y-glutamylcysteine synthetase with a plastid targeting sequence under control of the 35S promoter, display up to fivefold higher enzyme activity accompanied by an enhancement of PC2 and PC3 formation.

This resulted in an increase in Cd accumulation and Cd tolerance by about twofold as determined from root growth in the presence of Cd^{2+}. Similar effects were generated by the 35S-driven overexpression of an *E. coli* glutathione synthase (encoded by the *gsh2* gene) in the cytosol of *B. juncea* plants. Again, the higher activity of a glutathione biosynthetic enzyme led to an increase in PC formation, Cd tolerance, and Cd accumulation.

These results are interpreted as an indication that glutathione biosynthesis represents a rate-limiting step for PC synthesis. This view is supported by the basal constitutive expression of both PCS activity and PCS genes. However, there are indications for an up-regulation of PCS genes under metal stress, and the effects of PCS overexpression on Cd tolerance in plants have yet to be reported.

Also, a strategy to boost PC synthesis further could be to combine PCS overexpression with the overproduction of glutathione and/or cysteine.

These experiments will show whether PCS overexpression in plants-despite the apparent noninvolvement of phytochelatin formation in naturally selected hypertolerance (see later)-can be utilized for a significant enhancement of tolerance toward Cd and potentially other metals.

13.1.1.2 Metallothioneins

Metallothioneins (MTs) are a ubiquitous group of small cysteine-rich proteins. The class I MTs of mammalian cells are involved in cytosolic metal ion buffering. In yeast, MTs bind predominantly copper ions, whereas mammalian MTs bind mainly Zn.

The copper-inducible S. *cerevisiae CUP1* contributes to copper detoxification. A disruption of cupl causes Cu hypersensitivity; the overexpression of cup l. enhances Cu tolerance. CUP! as well as the mammalian MTs can also bind Cd.

MT Cd complexes are formed in S. *pombe* cells expressing the *S. cerevisiae* MT gene. For a number of mammalian cell lines it was shown that MT deficiency results in Cd hypersensitivity and that MT overexpression leads to an elevation of Cd tolerance. Still, in the light of numerous genetic experiments demonstrating the nonessentiality of MTs under normal growth conditions, the actual physiological function of MTs is considered enigmatic.

To date, more than 50 MT like sequences have been found in a variety of plant species. Plant metallothioneins are similar to class II MTs; i.e., they differ considerably from class I MTs in the alignment of Cys residues.

MT genes from *Arabidopsi s* complement MT deficient yeast and *Synechococcus* mutants, which are Cu and Zn hypersensitive, respectively, and they confer Cd tolerance to *S. cerevisiae* when overexpressed. It has been shown that MT genes are expressed in various plant tissues, and there are examples of induction by Cu. A correlation between MT expression and Cu tolerance has been observed for 10 different *Arabidopsis* ecotypes.

However, because genetic data and direct evidence for the binding of metals have been scarce, the role of metallothioneins in plant metal homeostasis is understood to an even lesser extent than in yeast and mammals. It remains a question whether MTs can function as metal chaperones analogous to ATX1 and others.

Metallothioneins from various sources have been constitutively expressed in plants, resulting in elevated Cd tolerance and changes in the Cd distribution between roots and shoots. The first such report dates back to 1987. A Chinese hamster MT II cDNA was inserted into a cauliflower mosaic virus. This recombinant virus was used to express metallothionein in infected *Brassica campestris* tissue, resulting in an increase in Cd-binding capacity and Cd tolerance.

Pan et al. introduced a synthetic gene coding for the a-domain of mouse MT I into tobacco. They were able to detect mouse metallothionein in leaves of transgenic plants and found significantly enhanced Cd tolerance. Similarly, human metallothionein was expressed in tobacco. Here, the authors analyzed the distribution of Cd in transgenic lines compared with the wild type and found that root-to-shoot translocation was reduced and the shoot content was 60-70% lower.

More recently, the transformation of *Nicotiana tabacum* with an MT gene from *Nicotiana plumbaginifolia* was reported (46). Again, the 35S-driven expression led to an elevation of Cd tolerance as determined by measuring growth and chlorophyll content. Data on the tolerance of the various MT-overexpres sing plants toward other metals have not been reported.

From the preceding resuits concerning MTs in yeast and mammals, one might expect to see some enhancement of Cu tolerance as well. The MT genes are also being exploited for enriching crops with micronutrients. In combination with genes encoding a phytase and ferritin (see later), an MT gene was introduced into rice to help increase the Fe content and improve Fe absorption in the human body.

13.1.1.3 Other Proteinaceous Chelators

In *S. cerevisiae,* Cu^+ ions are taken up by the high-affinity uptake systems CTR1 and CTR3 following reduction of Cu^{2+} by plasma membrane reductases. Once they are inside the cell, Cu^+ ions are bound by metallothioneins and at least three different chelators that serve specialized functions in copper delivery to different enzymes.

ATX1, originally identified as a suppressor of oxygen toxicity in sod⁻ strains, and its human homologue, HAH 1, direct Cu to a post-Golgi compartment via Cu-pumping ATPases. Yeast two-hybrid experiments provided evidence for direct interaction of ATX I with the yeast P-type ATPase CCC2.

Similarly, HAH 1 appears to bind to the Wilson disease adenosine triphosphatase (ATPase) depending on available copper, and the mutations in the transporter associated with the genetic defect impair Cu delivery to the ATPase. Other known Cu chelators in *S. cerevisiae* are COX17, involved in Cu delivery to mitochondria for insertion into cytochrome c oxidase, and LYS7, which functions as a Cu chaperone for the Cu,Zn-superoxide dismutase.

Both COX17 and LYS7 have human homologues. An ATX1 homologue was identified in *Arabidopsis* that complements an ATX1-deficient yeast strain, indicating the presence of similar chelation networks in plants. Furthermore, the *Arabidopsis* RANT represents a homologue of the Wilson disease ATPase. It is also suggested to be located in post-Golgi vesicles and to be involved in Cu delivery to the ethylene receptor.

Metal tolerance appears to be not directly related to this homeostatic Cu chaperone-chelator network. Yeast cells lacking one of the Cu chelators display a range of phenotypes ranging from oxygen radical susceptibility ($lys7^-$) to respiration deficiency ($coxl7^-$). However, no absolute requirement for Cu detoxification was observed. Instead, metallothioneins seem to play the key role as cytosolic Cu buffers in yeast cells.

The *Arabidopsis* ATX1 homologue is down-regulated under Cu stress. Nonetheless, it remains to be analyzed whether an increase in Cu binding capacity by overexpression of the Cu chelators would result in an enhancement of Cu tolerance. Other specific cytosolic chelator proteins involved in trafficking of essential metal ions await identification in plants.

Known and used biotechnologically is the Fe-binding protein ferritin, which can be found in bacteria, animals, and plants and functions in iron storage. Overexpression of a soybean ferritin gene specifically in rice seeds under the control of a glutelin promoter resulted in transgenic rice plants with a threefold higher seed iron content.

13.1.1.4 Organic Acids, Amino Acids

The reactivity of metal ions with S, N, and 0 renders organic and amino acids potential ligands. A considerable number of studies have been published on variations in the content of citric, oxalic, and malic acid in relation to metal tolerance. Computer modeling has been used to predict metal speciation in different compartments and dependent on pH, nutrient, and metal supply.

Unequivocal evidence, however, for a role of organic and amino acids in plant metal tolerance has to date been mainly confined to the histidine response in some Ni hyperaccumulators and the excretion of malate or citrate as a mechanism of Al tolerance.

Alyssum lesbiacum, a highly Ni-tolerant Ni hyperaccumulator, contains significantly more Ni in shoots than the nonhyperaccumulating relative A. *montanum* when grown in the presence of Ni. This was found to be correlated with a dramatic increase in histidine concentration of the xylem sap of *A. lesbiacum*, a response that was not observed for A.

montanum. Histidine is an efficient chelator of Ni, and it was found by X-ray absorption spectroscopy that most of the Ni in A. *lesbiacum* is bound to histidine.

The essential role of histidine for Ni hypertolerance was confirmed by the finding that spraying of A. *montanum* with histidine resulted in a doubling of biomass production under Ni stress. More recently, however, studies on another Ni hyperaccumulator, *Thlaspi goesingense*, showed that the histidine response appears not to be a general Ni tolerance mechanism.

Free histidine concentrations in shoot and xylem sap are no higher in *T. goesingense* than in the nontolerant *T. arvense*. Genes encoding histidine biosynthetic enzymes are not up-regulated upon Ni exposure.

Nonetheless, the effect of spraying histidine on A. *montanum* suggests that increasing the histidine content could enhance Ni tolerance, which is hypothesized to be the main determinant for Ni hyperaccumulation. It remains to be seen whether the efforts to generate His-overproducing plants by manipulating the histidine biosynthetic pathway will lead to the devel opment of Ni-hyperaccumulating transgenic plants.

Aluminum is the third most abundant element by percent weight in the outer earth's crust. Low pH in the soil enhances solubilization of bound Al. Because most plants are sensitive to micromolar concentrations of aluminum, Al toxicity is a major growth-inhibiting factor on acid soils.

Thus, the objective of engineering Al tolerance is to improve crop yields. Exclusion of Al from the root apex, which is the main site of Al uptake, has been established as a tolerance mechanism. In snapbean a correlation between Al-induced release of citrate and Al tolerance was observed, and for near-isogenic wheat lines differing in Al tolerance (Altl locus) it was shown that this trait cosegregates with high rates of malate release from the roots.

The protective effect of organic acid release is probably based on the fact that malate and citrate can chelate Al^{3+}, the presumed primary toxic Al species, and thereby reduce uptake of this ion into the symplast. These findings inspired a transgenic approach aiming at the production of Al-tolerant plants by increasing the citrate content.

A citrate synthase gene from *Pseudomonas aeruginosa* was overexpressed in tobacco and papaya under 35S control. The protein was targeted to the cytosol. Four tobacco lines with two- to threefold higher citrate synthase activity were analyzed. They contained up to 10-fold more citrate and released up to 4-fold more citrate into the medium.

Staining of root tissue with hematoxylin indicated that less Al is taken up by the citrate synthase overexpressors. In growth assays in the presence of Al, these lines showed a slight increase in tolerance. For instance, at 100 μM root growth was inhibited by 20-30% in the transgenic lines compared with about 50% in wild-type plants.

At 200 μM the inhibition was 40-60% instead of 85%. The benefits of higher citrate content were more pronounced for papaya plants. Transgenics with again two- to threefold higher citrate synthase activity developed roots in medium containing up to 300 μM Al, whereas control plants failed to develop roots in the presence of 50 pM Al.

The nonprotein amino acid nicotianamine, which is synthesized from S-adenosylmethionine and is a precursor of the phytosiderophore mugeneic acid, forms stable complexes with Fe(II). It may function as a transport form for Fe(II) and possibly other metal ions. The cloning of genes encoding nicotianamine synthases now makes it possible to study the role of this ubiquitous compound for metal ion mobility and regulation of metal responses in molecular detail.

A considerable number of metal chelators and chaperones remain to be identified. Little is known about the network governing the intracellular distribution of essential ions. The transport forms of metal ions in the xylem and phloem are not determined in sufficient detail. Furthermore, even the exact function of known chelators is largely unclear.

Are metallothioneins involved in Zn and Cu buffering, and do phytochelatins play a role in metal homeostasis? The Cd tolerance that is conferred by these compounds may well represent a product rather than their physiological function. Understanding and engineering the trafficking and translocation of metal ions have great potential for the generation of plants suitable for phytoremediation as well as of crops enriched in micronutrients.

13.1.2 Sequestration

Excess heavy metal ions are removed from the cytosol by sequestration. The compartment mainly involved in this process in yeast and plant cells is the vacuole. Mediators of sequestration are transporters in the respective membrane. The actual substrates for most of these transporters are as yet unknown.

13.1.2.1 Roles of ABC-type Transporters in Sequestration

Hmtl from *S. pombe* was the first gene identified that encodes a protein involved in vacuolar sequestration of heavy metals. *Hmt1* was cloned on the basis of the complementation of a Cd-hypersensitive *S. pombe* strain lacking vacuolar sulfide-containing PC-metal complexes. The corresponding protein belongs to the large family of ABC-type transporters. It localizes to the vacuolar membrane and transports PC-Cd complexes and apo-PCs.

S. pombe cells overexpressing *hmt1* accumulate more Cd^{2+} and are more Cd^{2+} tolerant. This indicates the potential of corresponding experiments in plants. However, no transgenic plants expressing *hmtl* have been described yet. Also, the respective ABC transporters that account for the transport of PC-metal complexes in plants are not yet known. The *Arabidopsis* MRPs 1-4 were shown to mediate the vacuolar sequestration of various xenobiotics conjugated to glutathione but lack PC transport activity.

This led to the speculation that additional factors might be involved in the sequestration of Cd ions as PC complexes in plants. In *S. cerevisiae,* the transporter YCF1 is required for Cd tolerance of yeast cells. YCFI-deficient cells are Cd hypersensitive. From transport studies with vacuolar membrane vesicles purified from wild-type and ycf *S. cerevisiae* cells it is known that YCF1 mediates the Mg-ATP-dependent transport of bis(glutathionato)cadmium into vacuoles.

YCFI apparently also represents one of two pathways in S. *cerevisiae* for arsenite detoxification. A ycf 1-knockout strain is As(III) hypersensitive and deficient in MgATP- and glutathione-dependent vacuolar uptake of "As(III).

13.1.2.2 Cation Diffusion Facilitators (CDFs)

The cation diffusion facilitator family constitutes another group of proteins involved in metal ion transport. First found in bacteria, members of this family have now been described in yeast, animals, and plants. The *S. cerevisiae* proteins COT 1 and ZRC 1 confer tolerance of cobalt (COT]) and zinc/cadmium (ZRC1) when overexpressed.

The observation that both proteins localize to the vacuolar membrane suggests a role in metal transport into the vacuole. *S. pombe* expresses one ZRC homologue that appears to be involved in Zn accumulation and Zn tolerance as a knockout strain is Zn hypersensitive and accumulates less Zn. Four mammalian zinc transporters (ZnT 1-4) of the CDF family are known.

ZnT 1 was cloned as a rat cDNA complementing the zinc sensitivity of a hamster cell line. The protein was detected in the plasma membrane and proposed to mediate Zn efflux. Both ZnT 2 and ZnT 3 reside in endomembranes, suggesting a function in zinc transport into the respective compartments: lysosomes (ZnT 2) and synaptic vesicles (ZnT 3) (80). To date, one CDF has been studied in *Arabidopsis,* at least two additional sequences can be found in the genome (Accession numbers AL353032, A0004561).

The *Arabidopsis* cDNA ZAT was isolated and, because sequence alignments indicated potential significance for metal transport, introduced into *Arabidopsis* in sense and antisense orientation fused to the 35S promoter. The sense lines showed a slight Zn tolerance phenotype when grown alongside control plants on medium containing toxic Zn concentrations.

No differences were observed in Cd sensitivity. When Zn accumulation was measured in a hydroponic culture, the sense line contained significantly more Zn in the roots than the control line, whereas shoot contents were similar for both lines. From these results the authors speculate that the ZAT protein might be a vacuolar transporter involved in the sequestration of Zn.

13.1.2.3 Other Transporters

Direct transport of Cd^{2+} ions could represent a pathway of Cd sequestration in addition to the transport of PC-Cd complexes. In tonoplast-enriched vesicles from oat roots, Salt and Wagner detected a Cd^{2+}/H^+ antiport activity. They demonstrated saturable ΔpH-dependent uptake of Cd^{2+} with an apparent K_M of 5.5 μM.

The authors estimated this K_M to lie in the range of cytoplasmic Cd^{2+} concentration for plants growing in Cd-contaminated soil, meaning that this antiport activity could play a role in Cd accumulation in the vacuole. Molecularly, the Cd^{2+}/H^+ antiport might be attributable to the same transporter as the Ca^{2+}/H^+ antiport in the vacuolar membrane. This was suggested by Salt and Wagner and others.

Ca^{2+}/H^+ antiporters CAX 1 and CAX2 have been cloned from *Arabidopsis* on the basis of the complementation of a Ca-hypersensitive *S. cerevisiae* mutant. CAX1 was shown to mediate high-affinity Ca^{2+}/H^+ exchange when expressed in yeast, and the K_M of CAX2 for Ca^{2+} is too high to be of any physiological relevance.

Instead, the authors mention for CAX2 high-affinity and highcapacity H^+/heavy metal cation antiporter activity. This would make CAX2 another target for engineering higher Cd accumulation in plants. No reports have been published yet on CAX2 overexpressors.

13.1.3 Efflux

The pumping of toxic heavy metal ions out of the cell represents the main tolerance mechanism in bacteria. Chromosomally or plasmid-encoded efflux systems have been found in all eubacterial groups studied so far. The metals for which specific transporters exist include copper, cadmium, zinc, silver, lead, and arsenite.

The respective transporter genes are generally part of metal tolerance operons also containing regulatory genes and genes coding for metal-binding proteins. Most of the bacterial metal transporters belong to the family of CPx-type ATPases, a subclass of the P-type ATPases. Well-studied examples are the Cd^{2+}-specific pump CadA from *Staphylococcus aureus* and the Cu pumps CopA and CopB from *Enterococcus hirae.*

Copper-transporting CPx-type ATPases have now also been characterized in eukaryotes. The human Wilson and Menkes disease proteins as well as the *S. cerevisiae CCC2* protein are involved in transport of Cu ions into a post-Golgi compartment. Defects in these systems lead to deficiencies in copper-dependent iron uptake systems.

In addition to functioning in the intracellular Cu distribution, eukaryotic CPx-type ATPases can mediate Cu tolerance in a fashion similar to that in the bacterial systems. In *Candida albicans* a Cu pump was identified that is essential for Cu tolerance and localizes primarily to the plasma membrane, suggesting an efflux pump activity.

The one functionally characterized CPx-type ATPase from plants, RAN 1 from *Arabidopsis,* probably serves a function similar to CCC2 and the Wilson and Menkes disease proteins. It is proposed to be involved in supplying Cu to the ethylene receptor. However, there are numerous other sequences in the *Arabidopsis* genome containing the signature motifs of CPx-type ATPases, such as the putative metal-binding CxxC elements in the amino terminus. At least two of the protein sequences show homology to the Cd^{2+} pump CadA from *S. aureus.*

Thus, there could well exist efflux systems in plants conferring heavy metal tolerance. To date, no such activity has been described physiologically in plants and the functional data have not been published for any of the other putative plant metal pumps. It is conceivable that these pumps, as well as the bacterial pumps, can be exploited for enhancing the tolerance of plants to Cu, Cd, and possibly other metals.

13.1.4 Engineering Metal Uptake Systems

We have seen tremendous advances in our molecular understanding of plant metal uptake. By complementation of respective yeast mutants, uptake transporters have been cloned for the micronutrients Fe, Cu, and Zn. Potential pathways for the entry of the nonessential metals Pb and

Cd have been identified. The cloning of transporters and the thorough characterization with regard to expression levels, tissue distribution, cellular localization, and biophysical parameters will make it possible to enhance plant metal tolerance or plant metal accumulation by altering the properties of plant metal uptake.

Conceivable strategies include the overexpression or the down-regulation of particular transporters as well as the expression of transporters with altered affinities and specificities.

13.1.4.1 The ZIP Family of Transporters

The ZIP family of metal ion transporters (for ZRT like, IRT like proteins) comprises proteins from eukaryotic organisms as diverse as trypanosomes and humans. The first member isolated was IRT1 from *Arabidopsis*. It complements the iron uptake deficiency of the S. *cerevisiae fer3fet4* mutant and encodes an Fe^{2+} transporter.

This activity and the expression in roots, which is inducible by iron limitation, indicate that IRT1 mediates uptake of Fe [21] from the soil. Additional experiments in yeast showed that IRT1 has a broad substrate range and also transports Mn^{2+}, Zn^{2+}, and, judging from competition experiments, Cd^{2+}.

The yeast homologues ZRTI and ZRT2 were identified on the basis of sequence similarities to IRT 1. They were subsequently shown to mediate high-affinity and low-affinity Zn^{2+} uptake, respectively. A zrtl/zrt2 mutant was then used to clone plant Zn uptake transporters. Three *Arabidopsis* cDNAs were found that complement the zinc uptake deficiency, designated ZIP 1-3.

They encode transporters with K_M values for Zn^{2+} in the nanomolar range, suggesting a role in plant Zn^{2+} uptake. Expression levels of ZIP transporters could represent one factor determining metal ion uptake and accumulation. When the Zn hyperaccumulator *Thlaspi caerulescens* and its nonaccumulating relative *T. arvense* were analyzed, it was found that Zn uptake into *T caerulescens* roots shows a K_M similar to that of the uptake in *T. arvense* but a 4.5-fold higher V_{max}.

Subsequently, a Zn^{2+} transporter was cloned from *T. caerulescens* that complemented the Zn uptake-deficient yeast strain zhy3 and displays kinetic properties matching the data for Zn uptake into *T caerulescens* roots. According to Northern analysis, this Zn transporter, termed ZNT1, is expressed at very high levels in *T caerulescens* irrespective of the Zn status whereas in *T arvense* the ZNT1 homologue mRNA is of low abundance in Zn-sufficient media and up-regulated under Zn-limiting conditions.

These observations suggest that by altering the expression levels of uptake systems it is possible to engineer metal accumulation in plants.

13.1.4.2 Nramp Metal Transporters

Genes encoding members of the *Nramp* family of integral membrane proteins were identified through very diverse genetic screens. *Nrampl* determines sensitivity to mycobacterium infection such as tuberculosis or leprosy and gave its name, natural resistance-associated macrophage protein, to the gene family.

Nramp homologous sequences have now been identified in bacteria, fungi, plants, and animals. It has been shown that the *Nramp* homologues SMF 1 in yeast and DCT 1 /Nramp2 in mammals mediate the uptake of a broad range of metals.

In plants, the ethylene insensitivity gene *EIN2,* which functions in signal transduction, contains an *Nramp* homologous domain but has no demonstrated metal transport function . *Arabidopsis* homologues of *Nramp* genes have been cloned and their function as metal transporters has been shown both in the yeast heterologous expression system and in planta.

Disruption of the *AtNramp3* gene leads to an increase in Cd^{2+} resistance, whereas overexpression of this gene confers increased Cd^{2+} sensitivity *in Arabidopsis.* These results point to a role of *AtNramps* in physiological Cd^{2+} transport and Cd^{2+} sensitivity in plants. *AtNramp3, AtNramp4,* and to some extent *AtNrampl* complement the phenotype of *fet3fet4*, a yeast mutant deficient in Fe uptake.

This result indicates that two families of Fe transporters, *IRTs* and *AtNramps* may contribute to Fe homeostasis in plants. The role of *AtNramps* in Fe transport in planta is supported by both the induction of *AtNramps* upon Fe starvation and the observation that AtNrainp3-overexpressing plants can accumulate higher levels of Fe upon Cd^{2+} treatment, whereas AtNrampl-overexpressing plants confer resistance to toxic levels of Fe.

Thus, the broad specificity of *Nramp* transporters may render them good target genes to engineer metal nutrient uptake and transport in plants as well as toxic metal sensitivity and accumulation in plants.

13.1.4.3 The Uptake of Nonessential Metal Ions

There is no known biological function for the highly toxic metals Pb and Cd. Thus, it is assumed that no transporters specific for these metals exist. Instead, Pb^{2+} and Cd^{2+} ions most likely enter the cytosol of plant cells via nonselective cation transporters. Candidate proteins have been described. The screening for wheat cDNAs that complement a K^+ uptake deficiency of *S. cerevisiae* strain CY162 led to the isolation of a low-affinity cation transporter named LCT1. Testing of effects on Cd^{2+} sensitivity of expressing yeast cells resulted in the observation that LCTI mediates Cd^{2+} uptake.

Further studies revealed that the physiological substrates for LCT 1 are probably Ca^{2+} and perhaps Na^+. Other potential pathways include some of the Nramp and ZIP transporters. The IRTI-mediated Fe^{2+}, Mn^{2+}, and Zn^{2+} uptake is inhibited by C^{2+}. For ZNTI, low-affinity Cd^{2+} uptake activity has been demonstrated directly. A cyclic nucleotide-gated channel from tobacco (NtCBP4) has been described as a first example of a plant transporter mediating Pb uptake.

Originally identified as a calmodulin-binding protein, this channel was found to localize to the plasma membrane. NtCBP4-overexpressing tobacco plants exhibit increased sensitivity toward Pb^{2+}, which correlates with enhanced Pb^{2+} accumulation. Interestingly, NtCBP4 overexpressors at the same time are more Ni^{2+} tolerant.

Possible explanations for this effect are, as suggested by the authors, interaction of NtCBP4 with Ni^{2+}, which attenuates uptake, or the suppression of other, more Ni^{2+} selective channels, by NtCBP4 overexpression. The dual effect described in this study provides another example of the potential of engineered changes in the expression of metal ion uptake systems for enhancing plant metal tolerance or accumulation. This approach will certainly be developed further with a more detailed understanding of the molecular structure of transporter proteins.

Elucidation of specificity and affinity determinants, for instance, will allow the expression of mutated proteins that exhibit fewer undesired activities. A more indirect way of potentially minimizing toxic metal uptake is related to iron nutrition. *S. cerevisiae* cells are rendered more metal sensitive by defects in the high-affinity Fe uptake system.

This is due to the Fe deficiency-induced expression of lowaffinity uptake transporters such as FET3, which are less selective and facilitate entry of other heavy metal ions. Similarly, Cd^{2+} uptake into root cells of Fe-deficient pea seedlings is sevenfold higher than into Fe-sufficient pea seedlings. Because the kinetic properties are not affected by the iron status, this difference is attributable to the induction of iron-regulated IRTI-like transporters. Thus, enhancing the iron efficiency of plants may lead to a reduction in undesired uptake activities.

13.1.5 Biotransformation and Direct Removal

Mercury is one of the most hazardous heavy metal contaminants worldwide and has caused numerous ecological disasters, e.g., the poisoning of Minamata Bay in Japan in the 1950s and 1960s. Toward the goal of developing plants suitable for the phytoremediation of Hg-contaminated sites, genes of the *mer* operon of gram-negative bacteria have been very successfully used.

MerA is a mercuric ion reductase catalyzing the production of elemental volatile Hg^0 from Hg^{2+}. Following extensive modification of the coding region for plant-optimized codon usage and insertion of a consensus plant translation signal into the 5' upstream region, the *merA* gene was first introduced into *Arabidopsis* under 35S control.

The resulting transgenic lines grew and developed in the presence of Hg^{2+} concentrations (50-100 μM) that wild-type plants cannot tolerate. In accordance with the observation that *E. coli* expressing merA are more resistant to Au^{3+} ions, the transgenic *Arabidopsis* lines also grew better than the wild type on Au^{3+} containing growth medium.

When Hg^0 evolution was assayed in several lines, it was found to correlate well with merA mRNA levels and mercury resistance. In order to evaluate the feasibility of this approach for phytoremediation, *merA* was also expressed in the forest tree yellow poplar *(Liriodendron tulipifera)*.

Three constructs differing in the extent of modification of the coding region were used to transform proembryonic masses by particle bombardment. Again, *mnerA* expression led to the generation of Hg^{2+} resistance. Transgenic colonies, embryos, and plantlets were able to grow in the presence of normally toxic $HgCl_2$ concentrations. The most extensively altered construct produced the highest number of Hg^R colonies. The Hg^{2+} resistance correlated with the presence of the MerA protein and Hg^0 release.

The principal form of mercury that biomagnifies, i.e., accumulates in ecosystems, and caused the poisoning of Minamata Bay is methyl mercury (CH_3Hg^+), which arises through bacterial methylation of Hg(II). It is about 100-fold more toxic than Hg(II), probably because it can easily diffuse through biological membranes.

The *mer* operon that was identified in bacteria isolated from mercury-contaminated environments consists of *merA*, a few other genes encoding Hg transporters and regulatory factors, and *merB*, which codes for an organomercurial lyase. This enzyme catalyzes the formation of Hg(II) from CH_3Hg^+ via protonolysis of the C—Hg bond.

In order to generate plants suitable for phytoremediation of methyl mercury, the *merB* gene was modified and expressed in *Arabidopsis,* analogous to *merA*. Transgenic plants transcribing *merB* mRNA and synthesizing MerB protein were tested for growth on phenylmercuric acetate and monomethylmercuric chloride.

At concentrations (0.5-2 AM) that are growth prohibiting for wild-type plants, the *merB* expressors germinated and grew well. The accumulation of Hg", which has to be assumed for plants containing an organomercurial lyase and growing in the presence of methyl mercury, is apparently tolerable because Hg(II) is far less toxic and phytochelatins are likely to sequester some of it.

Nonetheless, plants showing both the lyase and the reductase activity would represent even more efficient tools for the removal of mercury from contaminated sites. Consequently, the group of Richard Meagher crossed *Arabidopsis merA* and *merB* lines.

The respective plants were found to be up to 10-fold more tolerant to methyl mercury than transgenics expressing *merB* alone, demonstrating the potential of this approach. Selenium is an essential element that can be toxic when high levels are reached, for instance, in wetlands as a consequence of irrigation.

Because selenium is similar to sulfur, selenocysteine can be formed, which, when incorporated into proteins, may lead to dysfunction. Two principal pathways for detoxification can be found in plants, chemical reduction and incorporation into organic compounds. The uptake of Se occurs mainly as selenate via high-affinity sulfate transporters.

Because the reduction appeared to be the main factor limiting accumulation, the *Arabidopsis ATP* sulfurylase (APS), hypothesized to be the enzyme catalyzing selenate reduction, was overexpressed in *Brassica juncea*. The transgenic lines showed twofold higher APS activity and increases in selenate reduction, selenate accumulation, and Se tolerance.

13.1.6 Repair Mechanisms

One component of metal tolerance could be the repair of metal-induced damage. Copper is known to cause plasma membrane leakage resulting in K^+ efflux. A correlation was found in differentially Cu-sensitive *Arabidopsis* ecotypes between Cu sensitivity and the ability to reverse the K^+ efflux following Cu exposure. Also, lipid metabolism genes, possibly involved in membrane repair, are induced in *Arabidopsis* by Cu treatment.

Cadmium can denature proteins by interacting with SH groups and by replacing Zn. That the removal of abnormal proteins formed upon Cd exposure is of importance for the basic cellular cadmium tolerance was demonstrated for *S. cerevisiae*. Yeast cells with defects in either specific ubiquitin-conjugating enzymes or the proteasome are Cd hypersensitive. Moreover, expression of the ubiquitin system is induced by Cd treatment. This finding could not be confirmed for the *Arabidopsis* homologues. Thus, it remains unclear whether the UBCs play a similar role in plants.

13.2 MECHANISMS OF HYPERTOLERANCE

Species and genotypes, which are part of the heavy metal plant communities established on sites enriched in toxic heavy metals, are necessarily hypertolerant. The molecular mechanisms underlying this trait are little understood.

The typical plant response to metal exposure, the formation of phytochelatins, appears not to be involved in naturally occurring hypertolerance. *Silene vulgaris* ecotypes differing in Cu, Zn, or Cd tolerance have been investigated thoroughly in this respect.

When Cd-tolerant and -sensitive lines were compared, it was found that the sensitive lines form more PCs in their roots upon Cd exposure. Furthermore, synthesis of PCCd complexes is more rapid in sensitive lines and the PC composition is the same in sensitive and tolerant lines. Equivalent results were obtained for *S. vulgaris* ecotypes with differential Cu and Zn tolerance.

At Cu concentrations that were affecting tolerant and sensitive plants equally, the phytochelatin accumulation was also equal. The Zn-tolerant *Silene* plants produced less PCs than Zn-sensitive plants after exposure to both the same Zn concentration and equal-effect Zn concentrations.

The chain length distribution of PCs in both ecotypes was the same, indicating that there is no difference in PC stability explaining the differential tolerance. The more rapid PC synthesis in sensitive *Silene* plants is interpreted as an indication of higher Zn concentrations in the cytosol.

This difference is due not to a reduction of uptake but to increased compartmentalization of Zn in tolerant plants. There is evidence for enhanced Mg-ATP-energized vacuolar transport of Zn in tolerant ecotypes. This 2.5- to 3-fold difference cosegregates with Zn tolerance in crosses between Zn-tolerant and Zn-sensitive plants, which clearly shows the importance of vacuolar sequestration for Zn tolerance.

Together with the effects of Zn transporter overexpression in *Arabid-*

opsis, these data represent a promising lead for the generation of Zn-hypertolerant plants.

The other example of a clear correlation between metal hypertolerance and a specific response or mechanism is the previously mentioned Niinduced synthesis of histidine in some Ni hyperaccumulators, which results in efficient chelation of nickel.

13.3 CONCLUSIONS

Mainly owing to physiological studies on model species and the use of *S. cerevisiae* for the functional expression of plant genes, significant advances have been achieved in the molecular understanding of plant metal tolerance and homeostasis mechanisms.

A number of metal transporters involved in uptake and sequestration have been identified, and genes encoding enzymes involved in chelator synthesis have been cloned. The observations that Nihyperaccumulators synthesize more histidine and Zn-tolerant *Silene vulgaris* lines show enhanced vacuolar Zn uptake provide valuable leads for the elucidation of naturally selected plant metal tolerance.

Furthermore, using genes predominantly from bacterial sources, transgenic plants have been generated with enhanced mercury, methyl mercury, aluminum, and cadmium tolerance. However, we are only beginning to understand the complexity of metal tolerance and accumulation. Many fundamental questions remain to be answered.

For instance, we completely lack understanding of root-to-shoot translocation. Different plant species and even different genotypes of the same species vary greatly in the root/shoot ratio of metal accumulation. This ratio is the net result of the interplay of many different factors including uptake into root cells, sequestration in root cells, *efflux*, long-distance trans port, uptake into leaf cells, and storage.

Similarly, most of the players involved in the intracellular trafficking of metal ions remain unknown. Most likely, plant cells express specific chelator proteins such as the Cu chaperones known from *S. cerevisiae* that mediate the delivery of essential heavy metal ions to organelles and enzyme complexes.

These chelators interact with target proteins such as transporters localized in endomembranes. To date, RAN I is the only example known from plants. Also, clearly there is evidence for a sensing of metal status and for metal-dependent gene regulation. However, very little is known about underlying signal transduction chains and mediators of metal responses.

With the growing interest in plant metal homeostasis and the increasing number of researchers addressing these fundamental questions, the potential of biotechnological approaches to problems related to plant heavy metal tolerance will certainly evolve rapidly.

14

Tools and Techniques

It was not possible in the main text of this book to describe the experimental basis of all the conclusions presented. The objective of this appendix is to try partially to remedy this deficiency by describing some of the most important general methods used in the study of nucleic acids. For some of the methods it was necessary to provide a full description earlier, and in these cases only a brief reference is made here for completeness.

It should be stressed that the emphasis in this section is on the principles underlying the methods, and no attempt is made to provide laboratory protocols, which are now available in abundance, and to which reference will be made.

It is inevitable that some of the methods dealt with here (especially those involving cloning), although pertinent to the work already described in this text, will in many cases be superseded in the lifetime of this edition.

The reader wishing information about current methodologies is therefore advised to consult the most recent texts. However, at the time of writing, the cited references are recommended for more detailed information on principles and practice.

14.1 OCCURRENCE AND CHEMICAL ANALYSIS

14.1.1 Chemical Determination of Nucleic Acids in Tissues

The usual approach to the determination of nucleic acids in tissues involves fractionation to remove lipids and proteins etc., followed by assay on the basis of either phosphorus, ribose or deoxyribose, or purine and pyrimidine. The tissue is treated with cold dilute trichloroacetic or

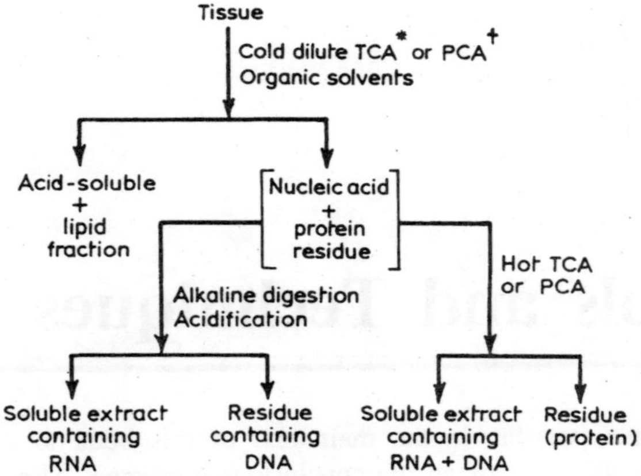

Figure 14.1: Extraction and fractionation of nucleic acids from tissues. TCA-trichloroacetic acid, PCA - perchloric acid.

perchloric acid to precipitate nucleic acid and protein, and lipid removed by extraction with organic solvent. In the procedure of Schneider total nucleic acid (DNA + RNA) can be separated from protein by hydrolysis in hot acid, the liberated products being soluble.

These latter are then assayed for ribose and deoxyribose. In the procedure of Schmidt and Thannhauser the DNA and RNA are separated by alkaline hydrolysis of the latter followed by acidification. Analysis of nucleic acid phosphorus in such fractions usually employs a colour reaction involving the formation of a phosphomolybdate complex.

Following depurination, deoxyribose can be determined by a colour reaction with diphenylamine. This reaction is specific for DNA. In contrast, the orcinol colour reaction employed for depurinated RNA also occurs to a lesser extent with DNA.

The spectral properties of the purine and pyrimidine bases of such hydrolysed nucleic acids may also be used as a basis of nucleic acid assay. These bases have absorption maxima in the region of 260 nm. The spectral properties of double-stranded and single-stranded DNA have already been discussed, as has the effect of pH thereon.

Protein also absorbs at 260 nm, although to a much lesser extent, dependent on the content of the aromatic amino acids phenylalanine, tryptophan and tyrosine. Because the latter two of these have an absorption maximum at 280 nm it is possible to obtain an approximate estimation of total nucleic acid and protein in a mixture by employing the following formulae based on :

$$\text{Nucleic acid (mg/ml)} = 0.064\ A_{260} - 0.031\ A_{280}$$

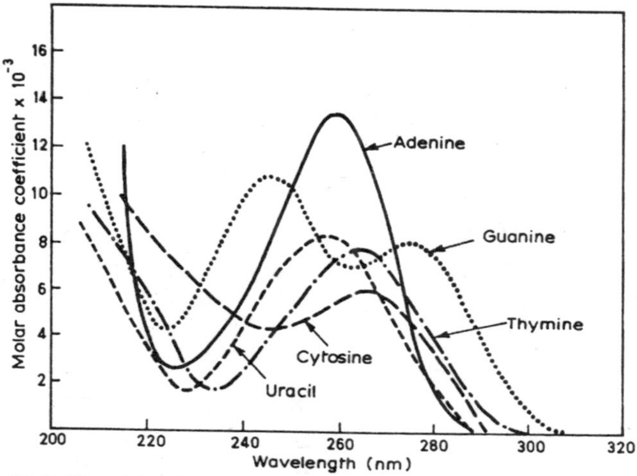

Figure 14.2: Ultraviolet absorption spectra for purine and pyrimidine bases at pH7.

$$\text{Protein (mg/ml)} = 1.45\ A_{280} - 0.74\ A_{260}$$

Typical values for the nucleic acid and protein contents of some rat tissues are given in Table elsewhere in this chapter

Table 14.1: Typical values for the concentrations of nucleic acids in different rat tissues.

Tissue	*RNA (mg/g protein)*	*DNA (mg/g protein)*	*RNA/DNA*
Brain	40	15	2.7
Kidney	43	18	2.4
Liver	54	12	4.5
Skeletal muscle	62	13	4.8
Spleen	58	58	1.0
Thymus	30	93	0.3

These may be expressed per gram tissue wet weight, but it is often useful to express them in terms of tissue protein, which may be easily estimated. Values of haploid DNA content for various organisms are given in Table elsewhere in this chapter.

14.1.2 Analysis of Base Composition and Nearest-neighbour Frequency

The molar proportions of bases in nucleic acids are determined by hydrolysis, separation and spectral quantitation, as described for DNA in other section of this chapter. Typical values for DNA are given in

Table 14.2: Molar percentages of bases in RNAs from various sources.

Type	*Species*	*Adenine*	*Guanine*	*Cytosine*	*Uracil*
Ribosomal					
E. coli	16S rRNA	25.2	31.6	22.9	20.3
	23S rRNA	26.2	31.4	22.0	20.4
	5S rRNA	19.2	34.2	30.0	16.7
Rat	18S rRNA	22.5	29.2	26.5	21.8
	28S rRNA	16.6	35.5	31.6	16.2
	5S rRNA	18.3	32.5	27.5	21.7
Messenger					
E. coli	Outer-membrane lipoprotein	28.3	24.2	22.7	24.8
Rabbit	β-globin	23.6	27.2	23.0	26.2
Rat	α-actin	21.7	26.4	30.6	21.3
Viral					
	Bacteriophage MS2	23.4	26.0	26.1	24.5
	Poliomyelitis virus	29.6	23.0	23.3	24.1
	Rous sarcoma virus	23.8	28.8	25.3	22.1
	Tobacco mosaic virus	29.1	24.2	19.1	27.6

Table elsewhere in this chapter and the variation in percentage GC content in different genomes can be found in Table elsewhere in this chapter. The principles of determining the base composition of RNA molecules are similar to those for DNA, and values for some types of RNA (actually based on sequence analysis) are given in Table elsewhere in this chapter.

Transfer RNA is not included because of the large proportion of modified bases, which, in the case of rRNA, have been ignored in Table elsewhere in this chapter. It may be remarked that although it was stated in the other chpater of this book that rRNA is GC-rich, this need only be true in relation to the overall base composition of the genome, rather than in absolute terms.

The examples of mRNAs in Table elsewhere in this chapter are rather arbitrary, and dramatic variations in the base compositions of different mRNAs are seen in the same organism.

The determination of nearest-neighbour frequencies (i.e. the relative frequency of occurrence of the sixteen different dinucleotides, NpM) in DNA, although to a considerable extent superseded by nucleotide sequence determination, still has applications.

14.1.3 Estimation of the Molecular Weight of DNA

Methods of estimating the size of small DNA molecules (up to 10 kbp) using *electrophoresis* on acrylamide and agarose gels are described in other section of this chapter. Such methods can be accurately calibrated using DNA molecules of known sequence. Although measurement of the *sedimentation rate* can be used to estimate the molecular weight of an RNA molecule, this method is not readily applicable to DNA molecules because of their extremely large size and asymmetric shape.

The theory behind the calculations depends on a knowledge of diffusion coefficients but these are equally difficult to obtain either by sedimentation analysis or optical mixing spectroscopy. However, a series of empirical formulae relating sedimentation coefficient ($S^{o}_{20,w}$) and molecular weight (M) have been derived by comparing the sedimentation rates of molecules of known size. For example, the equation for linear, doublestranded DNA is

$$S^{o}_{20,w} = 2.8 + 0.00834\ M^{0.479}$$

On sedimentation of DNA to *equilibrium* in density gradients (usually of CsCl) the width of the band of DNA is inversely proportional to the square root of the molecular weight. As with light scattering the result is dependent on the concentration of the DNA and a series of values must be obtained and extrapolated to zero concentration. This is the most widely used, fundamental method of estimating the size of DNA molecules and is applicable to molecules of up to 10^8 daltons (150 kbp).

Light scattering was widely used in the 1950s to measure molecular weights of DNA up to 5000 kbp, but only since the introduction of the laser and the production of machines to enable study of scattering at low angles has the method been extended to molecules of larger size. However, at low angles the scattering by dust particles becomes a major problem and with large molecules different regions of the same molecule serve as scattering centres.

The above methods all suffer from the disadvantage that they are badly affected by any heterogeneity in the size of the DNA molecules. It is extremely difficult to prepare high molecular-weight DNA intact from chromosomes and these methods tend to reflect the size of the smaller molecules.

The problem is even greater when attempts are made to relate molecular weight to *viscosity* as this involves subjecting the DNA solution to shear forces which tend to break long DNA molecules. However, Zimm developed a new, low-shear, viscometer consisting of a slowly rotating tube (the Cartesian diver) immersed in a DNA solution under pressure, and a relationship has been established between intrinsic viscosity (η) and molecular weight (M) of linear duplex DNA similar to that between sedimentation coefficient and molecular weight:

$$0.665 \log M = 2.863 + \log (\eta + 5)$$

A more important observation is that after the Zimm viscometer has been used to measure the size of very large DNA molecules the Cartesian diver rotates in the reverse direction for a while after the power has been switched off.

This *viscoelastic* effect is caused by the stretched DNA molecules relaxing to the unstressed configuration. The theory derived by Zimm to explain this effect has been used to measure the size of the DNA in *Drosophila* chromosomes showing the method to be applicable to DNA molecules of molecular weight up to 79×10^9 (over 10^5 kbp).

The method has the major advantage that it measures the size of the largest DNA molecules in the solution. Two 'visual' methods of measuring the size of DNA molecules involve electron microscopy and autoradiography. With *electron microscopy* the Kleinschmidt technique involves spreading the DNA on grids coated with polylysine or polystyrene-4-vinylpyridine.

A drop of DNA solution in IM ammonium acetate containing 0.1 mg/ml denatured cytochrome c is applied to the surface of a solution of 0.2M ammonium acetate. A film of cytochrome c spreads across the surface, binding to the DNA which is disentangled and spread out. A grid is touched to the surface and the sample is dehydrated in ethanol. It may be stained with uranyl acetate or by evaporation of platinum on to the surface.

In the latter process the grid is rotated so as to cause the metal to pile up against all surfaces of the DNA, which stands out from the grid. Double-stranded DNA is readily visualized by this technique but single-stranded DNA (or RNA) molecules require the presence of formamide to prevent their collapse.

Even so they appear as rather wispy threads and may be stretched out to different extents. The incorporation of proteins which bind strongly to single-stranded DNA (e.g. phage T4 gene 32 protein) causes these regions to assume a profile thicker even than the native DNA regions and they now have a regular mass per unit length.

The presence, on the same grid, of marker DNA molecules of known length (e.g. SV40 DNA in the relaxed circular form) allows the size of an unknown DNA molecule to be found by measuring the relative contour lengths on electron micrographs. This method can readily be used with DNA molecules of up to about 10^3 kbp, but larger molecules cannot be accommodated in a single frame.

In denaturation mapping double-stranded DNA is partially denatured (usually by alkali treatment) and the unpaired bases reacted with formaldehyde to prevent their reannealing. The duplex DNA molecule now exhibits single-stranded bubbles in the (A + T) rich regions to give a characteristic map.

When DNA is labelled *in vivo* with (^{3}H)thymidine and the cells lysed on a slide to release the intact DNA molecules their position can be visualized by *autoradiography*. The radioactive DNA on the slide is covered with a layer of photographic film which is activated by the β-particles produced by the decay of the tritium to produce black grains.

As with electron micrographs the size of the DNA can be obtained by measuring the contour length of the autoradiographic image. This method uses light microscopy and so the field size is much larger than with electron microscopy. Cairns applied it to the study of the *E. coli* chromosome, which is 3000 kbp long.

14.2. ISOLATION AND SEPARATION OF NUCLEIC ACIDS

14.2.1 Isolation and Separation of DNA

The main problems to be overcome in isolating high molecular weight chromosomal DNA from bacterial or eukaryotic cells are the removal of protein (especially deoxyribonucleases) and RNA, and the avoidance of mechanical shearing. For eukaryotic cells a combination of pronase orprotease K,a detergent and phenol extraction can be used directly to disrupt the cells and remove protein.

For the cells of Gram-negative bacteria such as *E. coli* gentle lysis can be effected using lysozyme. Phenol extraction is used widely in the isolation of both DNA and RNA, which are retained in the aqueous phase while denatured protein collects at the interface between this and the phenol phase, which contains the lipids.

RNA may be removed from the DNA by using pure pancreatic ribonuclease, or by isopycnic ultracentrifugation in a gradient of CsCl, which is, in any case, a useful step for further purification. In *isopycnic ultracentrifugation* a density gradient is established which encompasses the buoyant densities of the molecules to be separated.

Molecules will assume a position on the gradient corresponding to their own buoyant density, after which no further separation can occur.

In this method an equilibrium is established, and the method is sometimes referred to as 'equilibrium ultracentrifugation' to distinguish it from the rate-zonal method.

As already mentioned, the buoyant density of double-stranded DNA varies with the mole fraction of (G + C). In general, however, RNA has a much higher buoyant density (about 1.9 g/ml) than doublestranded DNA (about 1.7 g/ml), which, in turn, is higher than that of protein (about 1.3 g/ml).

Single-stranded DNA is of slightly higher buoyant density than double-stranded DNA (e.g. for a double-stranded DNA of buoyant density 1.703 g/ml, the value for the singlestranded form is 1.717 g/ml). Isopycnic ultracentrifugation is also used to separate small bacterial plasmids from chromosomal DNA.

The basis of separation in this case is not differing GC contents (these are, in general, similar) but the fact that the chromosomal DNA is linear (normal methods do not extract the large circular bacterial chromosome intact) whereas the plasmids are circular. This difference is exploited by the addition of saturating concentrations of the intercalating fluorescent dye, ethidium bromide.

The intercalation requires that the DNA strands be forced apart, with concomitant decrease in buoyant density and partial unwinding of the double helix. This latter process is hindered in the closed-circular plasmid molecules with the result that they bind less ethidium bromide and have a higher buoyant density than the linear chromosomal (and open-circular and linearized plasmid) molecules.

Before attempting to separate plasmid and chromosomal DNA it is advisable to enrich the preparation in plasmid DNA. The most frequently employed method for achieving this in preparative work is the use of an alkaline pH during lysis and extraction. This denatures the DNA, causing strand-separation in the case of linear molecules.

On neutralization the latter will not renature, whereas the closed-circular plasmid DNA will and can be separated from the insoluble denatured DNA. For small-scale isolation of plasmid DNA, a boiling step is usually used in place of the treatment with alkali.

Ethidium/CsCl gradients are also used in the purification of small animal viruses (e.g. SV40) which contain cyclic genomes. In this case the initial purification involves precipitation of most of the high molecular weight cellular DNA adsorbed onto a precipitate of sodium dodecyl sulphate to leave behind the so-called Hirt extract.

Although most preparations of chromosomal DNA consist of sheared fragments, these do not in general separate on isopycnic ultracentrifugation because the fragmentation is random and the base compositions of individual fragments are generally similar. Exceptions do occur in the case of regions of repetitive DNA in eukaryotes which may be generated as quite homogeneous species in relatively large amounts if the fragments are of the appropriate size.

Should the base composition of such a repeated DNA be markedly different from the average base composition of the chromosomal DNA, resolution from the bulk of the latter will occur. Such DNA was originally given the descriptive name 'satellite DNA', a designation which is sometimes applied to all repetitive DNA, regardless of base-composition or function.

Such loose terminology is to be discouraged. It is also possible to isolate the tandemly repeated rDNA and histone genes of certain organisms by this method. These exceptions aside, separation of DNA by isopycnic ultracentrifugation is largely confined to the different chromosomal species one might find in a given cell. Precise fragments of such DNAs can be generated by restriction endonucleases and the separation of such fragments is a common requirement in recombinant DNA manipulations.

Most frequently this is achieved by horizontal *agarose gel electrophoresis*. This separates different linear DNA molecules according to size (the mobility is inversely proportional to the $\log_{10}$ of the molecular weight), as the charge per unit length of different molecules is identical and the larger molecules will be retarded by the gel matrix.

The percentage of agarose can be increased to separate smaller-sized DNA fragments, but for very small species *polyacrylamide gel electrophoresis* (which is often used preparatively for fragments up to 1 kb) must be used. This method is also employed in rapid nucleic acid sequencing methods, in which case 8M urea is included to ensure denaturation.

As agarose gel electrophoresis is frequently used to analyse preparations of plasmid DNA which may be slightly nicked, it is worth mentioning that the closed-circular supercoiled form of the plasmid migrates more rapidly than the less compact relaxed-circular form produced by single-stranded nicks. Any linearized plasmid molecules generally migrate at an intermediate position between these two, a result that is perhaps a little unexpected.

Nucleic acids may be precipitated with ethanol, 2.5 volumes in the presence of 0.3 M sodium acetate, being frequently employed.

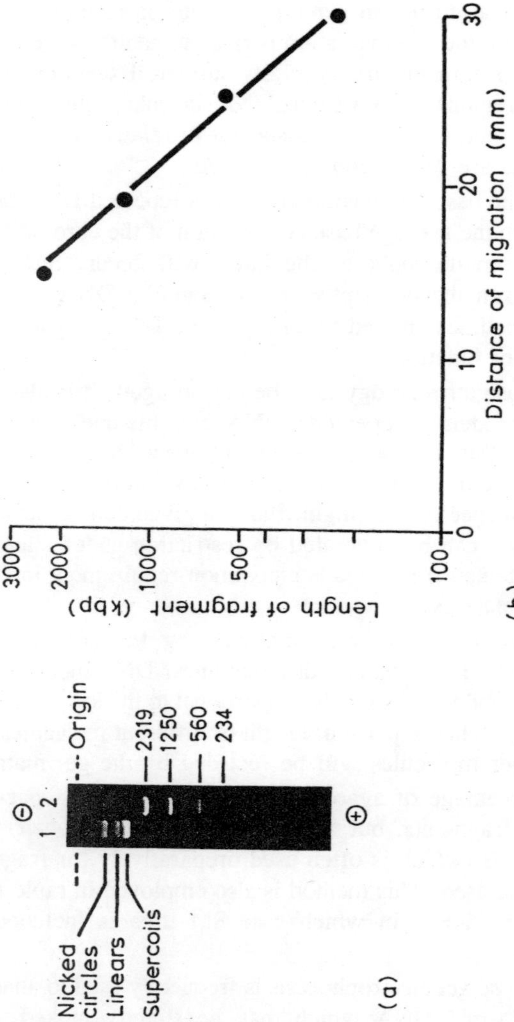

Figure 14.3: Agarose gel electrophoresis of DNA.

14.2.2 Isolation and Separation of RNA

In principle the methods used to extract RNA from cells are generally similar to those for DNA. Degradation of RNA by ribonuclease is a greater threat than that of deoxyribonuclease degradation of DNA, especially in the case of mRNA, but there are a variety of methods to counter this.

The major separation problem is posed by the large number of different RNA species in eukaryotes, and to study a particular species it is necessary to separate it from the others. One preliminary that may be useful in particular cases is *subcellular fractionation*.

Thus the nuclei may be removed by low-speed centrifugation (700 g for 5 min) after gentle disruption of the cells in the presence of sucrose (to preserve the osmolarity), the mitochondria may be removed by further centrifugation (10 000 g for 10 min), and the ribosomes separated from soluble RNA species such as tRNA by ultracentrifugation (e.g. 100 000 g for 90 min).

It may of course be necessary further to purify the subcellular fraction in which one is interested, and methods for nuclei, nucleoli, mitochondria, and ribosomes are available. It may be remarked that the most appropriate method for isolating an organelle, such as the nucleus, for subsequent RNA extraction (where purity is of greatest importance) may be quite different from the method which will yield the most biologically active material.

One of the most widely used methods of separating RNA molecules is *rate-zonal ultracentrifugation* employing *sucrose density gradients*. In this method the separation of RNA is based primarily on size (in principle, molecular shape can also have an influence, but with RNAs differences in shape make no significant contribution to the separation).

It should be emphasized that the density range of sucrose solutions does not extend to that of RNA species and, in contrast to isopycnic ultracentrifugation involving CsCl gradients, no equilibrium is reached. The main purpose of the sucrose gradient is to prevent diffusion of the zones of individual species that are separated from the narrow zone applied initially to the top of the gradient.

Isopycnic ultracentrifugation in gradients of dense sucrose (often employing a discontinuous or 'step gradient') may be used to separate 'free' ribosomes from those bound to the membrane of the endoplasmic reticulum.

Rate-zonal ultracentrifugation in sucrose density gradients is also used to separate different size classes of polyribosomes and to separate the large and small subunits of ribosomes after their dissociation. Rate-zonal ultracentrifugation is clearly inappropriate for the separation of different species of *tRNA,* which are broadly similar in size.

Methods involving separation on the basis of charge and hydrophobicity differences (e.g. chromatography on BD-cellulose and RPC5) may be employed for this purpose. The major problem to be overcome in the isolation of *mRNA* is purification from other species of RNA,

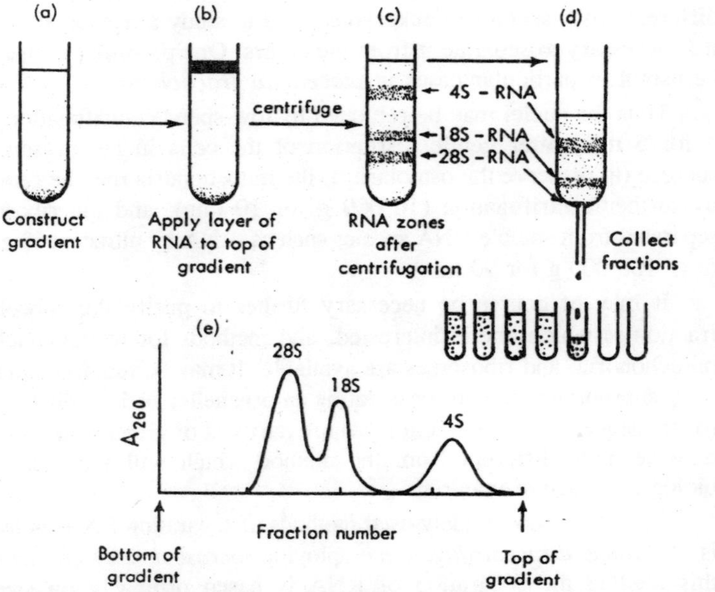

Figure 14.4: Rate zonal centrifugation of RNA through a sucrose density gradient.

which are approximately 20 times more abundant. Affinity chromatography using oligo(dT)-cellulose or poly(U)Sepharose—materials that bind the poly(A) tails of mRNAs—is the basis of the separation of mRNA from other species of RNA.

The minimum size of poly(A) tail required for binding to poly(U)-Sepharose is smaller than that for oligo(dT)-cellulose (about 6-10 residues compared with 15 residues), and although the latter material is in more widespread use, the former has applications to mRNAs having very short poly(A) tails.

A mixture of different mRNAs can be separated to a certain extent on the basis of size, either by sucrose density gradient centrifugation or, on a smaller scale, by agarose gel electrophoresis—basically as for DNA but generally under denaturing conditions.

Both agarose gel electrophoresis and polyacrylamide gel electrophoresis have been applied to other types of RNA. The isolation of individual mRNA species is usually achieved by recombinant DNA techniques.

14.3 HYBRIDIZATION OF NUCLEIC ACIDS

The denaturation and renaturation of homoduplex DNA molecules has been dealt with in other section of this chapter, where the effects of base composition, temperature and ionic strength were mentioned. The application of DNA renaturation to determining the copy number of different portions of eukaryotic genomes *(C_0t value analysis)* has been described in other section of this chapter.

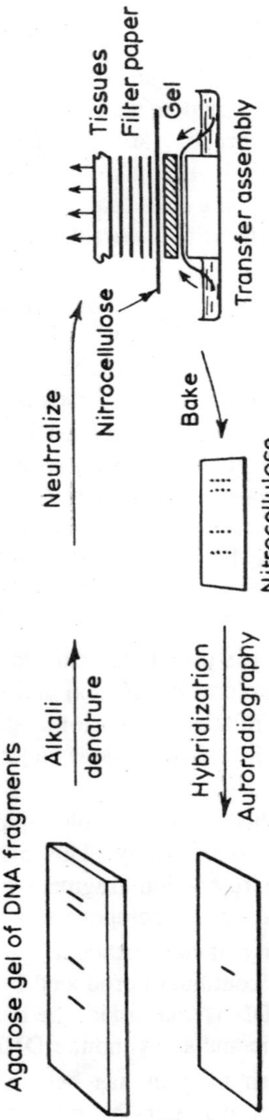

Figure 14.5: Southern blotting.

An important technique in molecular biology is the formation of heteroduplexes between different DNA molecules or between DNA and RNA. The application of electron microscopic visualization of such heteroduplexes formed in solution to the determination of the positions of introns *(R-loop mapping)* has already been described, and the electron microscopic analysis of heteroduplexes can be most useful in a variety of other situations.

However, for most routine work heteroduplexes are detected by the radioactivity of one of their components (methods for the radioactive labelling of nucleic acids—probes—are described elsewhere in this chapter). It is operationally convenient to have one component of the hybridizing pair immobilized on a solid 'paper' support in order that the hybrids may be easily visualized by autoradiography.

Before the introduction of heteroduplex formation involving immobilized nucleic acids analysis involved the time-consuming inconvenience of multiple solution-hybridizations, when the experimental protocol involved identifying which component of DNA fractionated by agarose gel electrophoresis hybridized to a particular radioactively-labelled probe.

To overcome this problem Southern devised a method of transferring the DNA from the fragile gel to nitrocellulose paper by means of capillary action. The DNA must be denatured with alkali before transfer, and afterwards it is baked at 80°C (in a vacuum oven to prevent it igniting) in order to fix the DNA permanently to the nitrocellulose. This method of transfer is commonly called *Southern blotting,* after its originator.

The nitrocellulose paper is usually hybridized with the radioactive DNA probe in a minimum volume of solution in a sealed polythene bag at 68°C and conditions of relatively high ionic strength that promote heteroduplex formation. (If 50% formamide is included the hybridization is performed at 42°C.)

The membrane is washed under suitable conditions (see below), dried, and subjected to autoradiography. This method is most easily applied to detecting which restriction fragment of a larger piece of cloned DNA contains sequences homologous to a particular probe.

However, the sensitivity of the method is such that it can detect individual fragments in the continuum produced by restriction enzyme digestion of total genomic DNA, and in fact the method was originally applied to uncloned genomic and sub-genomic DNAs.

Such genomic Southern blotting has become a routine tool in screening for human genetic disorders by the detection of restriction fragment length polymorphisms, for full details of which the reader is directed elsewhere.

The conditions of washing nitrocellulose membranes (nylon membranes may also be used) depend on the degree of homology between the two members of the heteroduplex. (The homology need not be 100% as hybridization may be between a variety of related but non-identical sequences.)

When the two members of the heteroduplex are identical the washing is usually performed 12°C below the melting temperature (T_m), which in a solution 0.2 M in Na^+ is related to the mole percentage of (G + C) by the equation:

$$T_m = 69.3 + 0.41\ (G + C)$$

The effect of ionic strength (p.) on the melting temperature is given by the equation:

$$T_{m2} - T_{m1} = 18.5 \log_{10} \frac{\mu_2}{\mu_1}$$

Thus, to allow for the fact that the Tm of duplex DNA decreases by 1°C per 1-1.5% mismatch the 'stringency' of the hybridization can be decreased by lowering the temperature of the wash and/or by increasing the ionic strength of the washing buffer.

Southern's method for the transfer of DNA from gels to paper has been extended to RNA, where it has acquired the somewhat ridiculous name of *Northern blotting*. (There is a third point on the compass, 'Western blotting', the name sometimes given to the electrophoretic transfer of proteins from polyacrylamide gels to nitrocellulose.)

The methodology of Northern blotting is different from that of Southern blotting because RNA does not bind to nitrocellulose paper under conditions in which DNA does, and is hydrolysed by alkali.

To overcome this problem, transfer can be made to diazobenzyloxymethyl (DBM) paper, which binds RNA (and DNA). RNA can, in fact, be transferred to nitrocellulose paper from agarose gels provided that it is denatured (e.g. by formaldehyde).

Hybridization of immobilized nucleic acids on paper supports is not restricted to material transferred from agarose gels. It can be used for detecting recombinant DNA in bacterial colonies or bacteriophage plaques after transfer from petri dishes or for multiple samples of total cellular RNA applied directly to the nitrocellulose (the so-called *Dotblot* technique).

14.4 METHODS OF LABELLING NUCLEIC ACIDS

There is a variety of circumstances in which the detection of DNA is only possible if the DNA is labelled in some way. Furthermore, certain techniques require DNA labelled specifically at one end. By far the most common method of achieving this is with radioactivity (usually ^{32}P; less commonly ^{35}S or ^{3}H), and the discussion below will be exclusively in terms of, this.

However, non-radioactive labelling methods are available. Currently the most prominent of these involves the use of biotin-labelled derivatives of dNTPs, which are detected by using an anti-biotin antibody. A variety of systems has been devised to visualize this, for example with a second antibody coupled to an enzyme such as horseradish peroxidase that catalyses a colour reaction.

14.4.1 General Labelling Methods

When the objective of labelling a nucleic acid is merely that of allowing its detection, methods that cause the incorporation of radioactivity throughout the molecule are suitable, as well as the specific end-labelling methods described elsewhere in this chapter. Historically, RNA and DNA were labelled from (^{3}H), (^{14}C) or (^{32}P)-labelled precursors

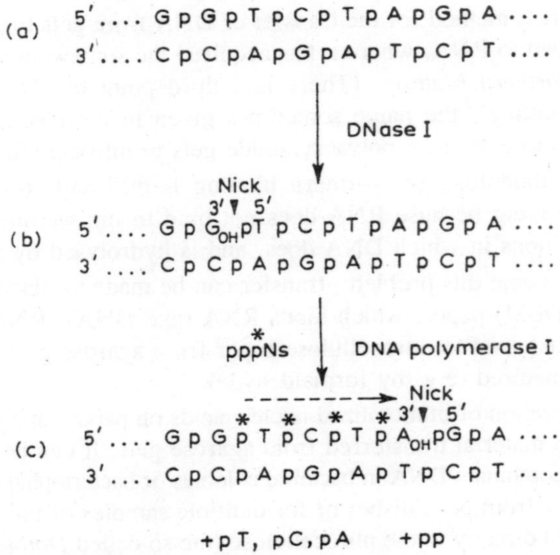

Figure 14.6: Nick translation.

in vivo, but methods involving the incorporation of radioactive label *in vitro* are more convenient for recombinant DNA work and produce material of higher specific activity. The most widespread method of labelling double-stranded DNA is by the method of *'nick translation'* using a(^{33}P)dNTPs.

In this method nicks are introduced into the DNA with DNase and the 5'-phosphate ends produced can serve as substrates for the 5' to 3' exonucleolytic activity of *E. coli* DNA polymerase I, which at the same time repairs the gaps by addition of the a(^{33}P)dNTPs to the 3'-OH end of the 'nick'.

Another replacement synthesis method involving T4 DNA polymerase has certain advantages over nick-translation, but is less widely used. Single-stranded labelled DNA probes may be prepared by copying the DNA after cloning into the single-stranded phage vector M13.

More recently a method of producing highly radioactively-labelled RNA copies of a piece of DNA has been introduced, which involves transcription with bacteriophage SP6 RNA polymerase from the DNA cloned into a special vector that puts it under the control of a bacteriophage SP6 promoter. Another method of producing very high specific activity DNA probes also involves copying but does not require preliminary subcloning.

The Klenow fragment of DNA polymerase is employed and random oligonucleotides are used as primers with the denatured DNA as template. It is possible to label mRNA directly *in vitro* by certain end-labelling methods but a copying method is generally preferred because it produces a more stable product of higher specific activity.

This involves the use of α (^{32}P)dNTPs and reverse-transcriptase to produce α (^{32}P)-labelled single-stranded cDNA copy. The labelling of tRNA *in vitro* is important in sequence determination in cases where insufficient material is available for spectral analysis and it is not possible to label with ^{32}*P in vivo.*

As well as enzymic end-labelling methods chemical labelling using (^{3}H)borohydride has been employed.

14.4.2 End-labelling Methods

These are particularly important in preparing DNA for sequence analysis by the method of Maxam and Gilbert and for certain other analytical purposes. For these purposes the label must be covalently incorporated and the products must be homogeneous.

The DNA to be labelled is generated by cleavage with a restriction endonuclease and the method of labelling depends on whether flush ends, or 'sticky' ends (in which there is a 5'-P or 3'-OH overhang) are produced. Ends with a 5'-P overhang are easily labelled by either of two methods, in one of which the 5'-P is removed by alkaline phosphatase and replaced by the γ-phosphate of γ(^{32}P)ATP in a reaction catalysed by bacteriophage-T4 polynucleotide kinase. In the other the Klenow fragment of *E. coli* DNA polymerase I is usually used to fill the gap in the strand with the recessed 3'-OH end, using an appropriate α(^{32}P)dNTP, although bacteriophage T4 DNA polymerase may also be used. (The 5' to 3' exonuclease of the complete *E. coli* enzyme would result in degradation of the template for this reaction.) Thus these two methods complement each other in labelling different strands.

The polynucleotide kinase method may be adapted for labelling the 5'-P of flush ends by effecting local denaturation of the latter, but cannot be applied efficiently where there is a 3'-OH overhang. The 3' to 5' exonuclease and 5' to 3' polymerase activities of the Klenow fragment of DNA polymerase I can be used to label flush ends or 3'-OH protruding ends by a replacement synthesis method, although bacteriophage T4 DNA polymerase (which has a more powerful 3' to 5' exonuclease activity and also no 5' to 3 exonuclease) is more effective.

Currently, the most effective method of end-labelling 3'-OH overhangs for sequencing involves the use of terminal transferase, but with α(^{33}P)2',3'dideoxy ATP to ensure addition of only a single nucleotide. A number of methods for end-labelling are available when the nucleic acid is only to be used as a probe. With DNA, extension of 3'-overhanging ends using terminal transferase and α(^{32}P)dNTP is an alternative to the use of DNA polymerase.

For mRNA, extension of the 3'-end with *E. coli* poly(A) polymerase can be used for labelling with α(^{32}P)ATP, or (^{32}P) cordecypin (3'-deoxy ATP) if the addition of only a single nucleotide is required. An alternative to covalent incorporation as a method of end-labelling, useful in certain restricted circumstances, is hybridization of a labelled fragment to suitable single-stranded regions. In the case of eukaryotic mRNAs, (^{3}H)poly(U) can be hybridized to the poly(A) tails.

14.5 DETERMINATION OF NUCLEIC SEQUENCES

14.5.1 Determination of DNA Sequences

There are two main methods currently in use for sequencing large fragments of DNA. These methods (that of Maxam and Gilbert and that of Sanger) both involve the generation of a ladder of fragments of

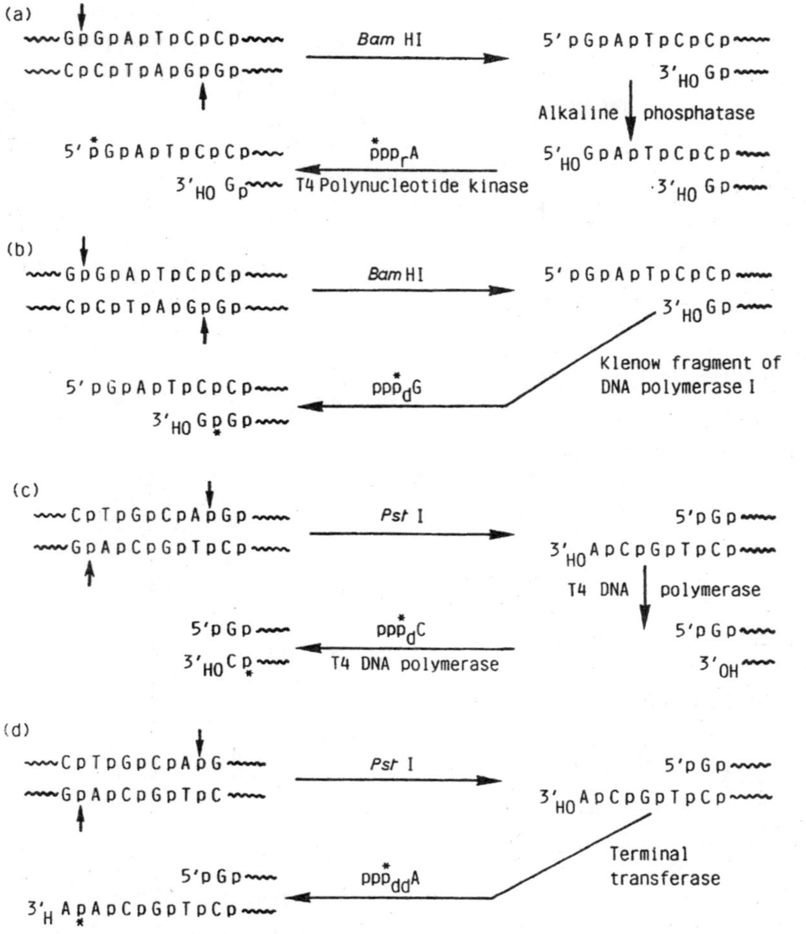

Figure 14.7: Methods of end-labelling DNA fragments.

different sizes but with one common end.

The basis of these methods is to generate specific sets of radioactively labelled fragments, each set terminating at a particular base (in the ideal case), and the use of high-resolution polyacrylamide gels allows fragments differing by only a single nucleotide to be resolved and the sequence to be deduced. In both cases cloned DNA is normally used, although the direct application of the Maxam and Gilbert method to genomic DNA has been described.

A third method of sequencing (the 'Wandering Spot' or 'Mobility

Shift' method), although no longer used for large DNA molecules, still finds application to small oligonucleotides (especially those produced synthetically), which are difficult to sequence by the other methods. An outline of earlier methods, now superseded, can be found in.

14.5.1.1 The Method of Maxam and Gilbert

The essence of this method is the chemical fragmentation of either single- or doublestranded DNA by base-specific reactions. The reactions most commonly used are those absolutely specific for guanine residues and for cytosine residues respectively, and those that are specific only for purines or pyrimidines, respectively.

Cleavage at guanine residues is effected by methylation with dimethyl sulphate at the N7 position, leading to instability of the glycosidic linkage which is then hydrolysed by piperidine, followed by, β-elimination of both phosphates from the sugar.

Purine nucleotide linkages are hydrolysed with acid, again followed by piperidine treatment. Pyrimidine residues are hydrolysed by hydrazine, a reaction which, in the case of thymine, can be inhibited by 2 M NaCl, thus allowing specific cleavage at cytosine residues. (Under the conditions normally used, cleavage does not occur at 5-methylcytosine residues, causing a gap in the ladder.)

Conditions of chemical cleavage are generally adjusted to try to obtain a single scission per DNA molecule. Even so, each scission would produce two fragments. In order to visualize on polyacrylamide gels only fragments of increasing length emanating from one end of the DNA, a single end of the DNA must be radioactively labelled using one of the methods described elsewhere in this chapter.

Such methods, in fact, generally label both ends of a piece of duplex DNA, and fragments with a single-labelled end are generated by restriction endonuclease digestion of this labelled DNA followed by separation on polyacrylamide gels. Less commonly, single endlabelled molecules are obtained by strand separation of the DNA. An example of a Maxam-Gilbert sequencing gel and its interpretation is given in Figure elsewhere in this chapter.

As sequence data accumulate, the use of computer programs to handle them and identify restriction endonuclease sites greatly facilitates sequencing by the Maxam-Gilbert method.

14.5.1.2 The Sanger Dideoxy Method

The essence of this method is the primed synthesis of partial copies of the DNA to be sequenced, with random base-specific premature

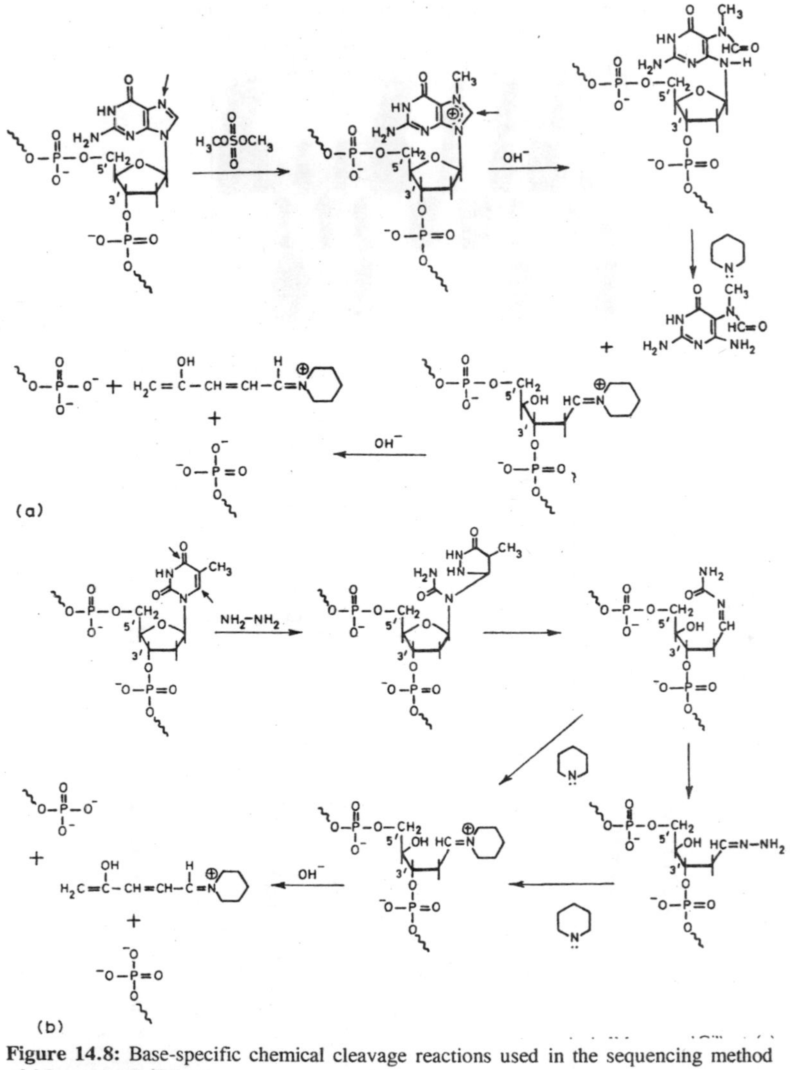

Figure 14.8: Base-specific chemical cleavage reactions used in the sequencing method of Maxam and Gilbert.

termination of the copying producing ladders of different length fragments terminating in each of the four different bases. The copying is catalysed by the Klenow fragment of *E. coli* DNA polymerase I, which requires a primer and that the DNA to be copied be in the single-stranded form. (The Klenow fragment, lacking the 5' to 3' exonuclease, is used to

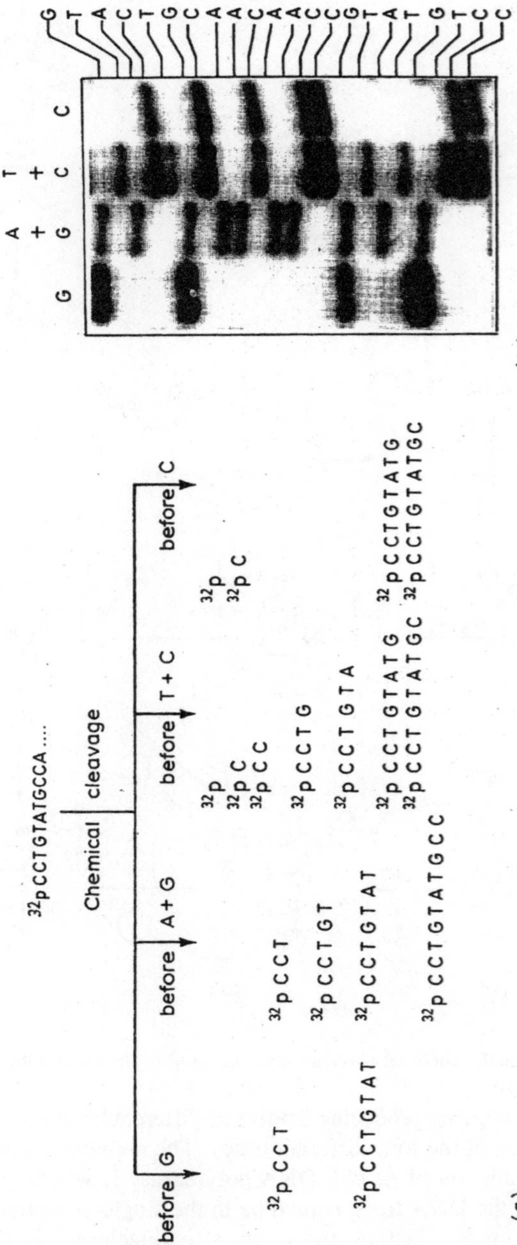

Figure 14.9: Example of DNA sequence determined by the method of Maxam and Gilbert.

prevent attack on the 5'-end of the primer.) The copies are made radioactive by inclusion of an $\alpha(^{32}P)$dNTP or a $\alpha(^{35}S)$dNTP in the reactions, and base-specific chain termination is effected by the addition of the appropriate 2'3'dideoxynucleotide triphosphate (ddNTP), which, lacking a 3'-OH, cannot be extended further.

An example of a Sanger sequencing gel and its interpretation is given in Figure elsewhere in this chapter. Initially, the widespread application of this method was prevented by the requirement for a single-stranded template and the need for a separate oligonucleotide primer for each piece of DNA to be sequenced.

However, these limitations were overcome when the single-stranded bacteriophage, M13, was adapted for use as a sequencing vector. The DNA to be sequenced may now be subcloned into one of a variety of adjacent sites in the double-stranded replicative form of the bacteriophage vector, and can be sequenced in the easily prepared singlestranded DNA using oligonucleotide primers complementary to regions of the bacteriophage DNA flanking the cloning sites.

This development has led to a dramatic increase in the use of the Sanger method. The random subcloning of fragments from the DNA to be sequenced has resulted in the need for computer programs to order the sequences obtained.

14.5.1.3 The Wandering-spot Method

The Maxam-Gilbert and Sanger methods described above, although extremely powerful for large restriction fragments of DNA, are not well suited to smaller oligodeoxynucleotides. The sequence of these can be determined following end-labelling by the chromatographic separation of a non-base-specific partial enzymic digest of the oligodeoxynucleotide in which all the partial digestion products having the same labelled end are represented.

Digestion is by venom or spleen phosphodiesterase and two-dimensional thin-layer homochromatography on DEAE-cellulose is employed. The shortening of the oligodeoxynucleotide by a single nucleot-ide causes a change in mobility that is characteristic of the nucleotide removed and which allows the sequence to be read.

14.5.2 Determination of RNA Sequences

Before the advent of rapid DNA sequencing methods considerable effort was expended on developing techniques for sequencing RNA. The availability of partially or totally basespecific endonucleases allowed employment of approaches formally rather similar to those used for

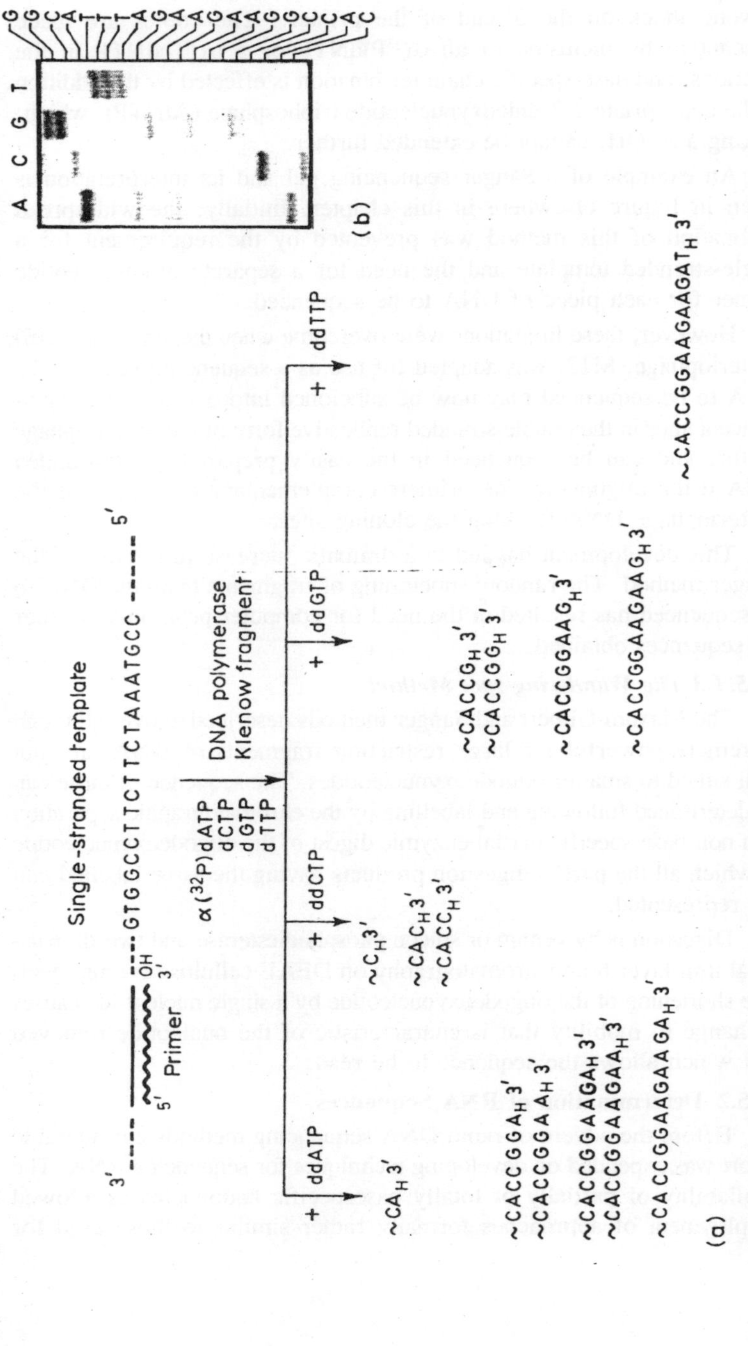

Figure 14.10: Example of DNA sequence determined by the dideoxy method of Sanger.

sequencing proteins, but requiring different methods for separating and analysing the oligoribonucleotides.

Although the sequencing of a DNA copy must now be the method of choice for sequencing any RNA, this approach may not be applicable to small RNAs or oligonucleotides, and in any case does not identify modified nucleotides. In these cases (especially for tRNAs) direct methods (including the 'wandering-spot' method, mentioned above) are still employed.

Rapid sequencing methods involving basespecific cleavage and separation of fragments on polyacrylamide gels have been introduced for 5'-end-labelled RNA. The cleavage is effected enzymically with ribonucleases, the following reactions being currently in use:

Ribonuclease T_1 : cleaves 3' of G
Ribonuclease U_2 : cleaves 3' of A
Ribonuclease Phy M : cleaves 3' of A and U
B. cereus ribonuclease : cleaves 3' of U and C

In other respects the method and the interpretation of the results are formally similar to that of Maxam and Gilbert, described elsewhere in this chapter.

14.6 RESTRICTION MAPPING OF DNA

When studying specific regions of genomic or cloned DNA it is necessary to have some sort of 'map' to differentiate one area from

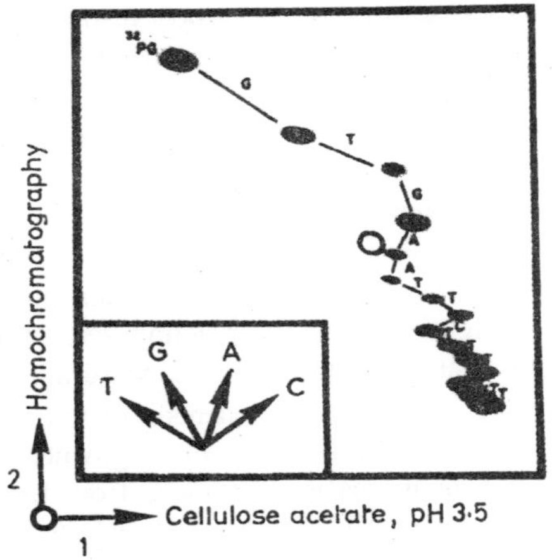

Figure 14.11: Example of oligodeoxynucleotide sequence determined by the 'wandering spot' method.

another. There are two different types of map. *A genetic map* can be constructed (in suitable organisms) in which the positions of functional genes contributing to particular phenotypes can be interrelated by studying the frequency of genetic recombination between normal and mutated alleles.

A *physical map,* on the other hand, can be constructed in the absence of genetic data or knowledge of any function the DNA might have, and uses purely physical techniques. These include the formation of heteroduplexes between standard DNA molecules and the DNA under investigation, and restriction mapping, the subject of this section.

A restriction map indicates the positions along a piece of DNA at which there are recognition sites for particular type II restriction endonucleases. The method in its simplest form involves digesting the DNA separately and in combination with different restriction endonucleases, separating the resulting fragments by agarose (or sometimes polyacrylamide) gel electrophoresis, visualizing these under ultraviolet light after staining with ethidium bromide, and estimating their size in relation to standards.

If there are not too many recognition sites for the enzymes it is possible to deduce the position of these as illustrated in Figure elsewhere in this chapter. It can be appreciated that for enzymes with many recognition sites it is both difficult and tedious to construct a restriction map by this procedure.

In the case of large pieces of DNA (e.g. those cloned in bacteriophage lambda or cosmid vectors) most restriction endonucleases will have many recognition sites and it is difficult to circumvent this problem by choice of enzyme. In any case it is often useful to have a restriction map for enzymes that cleave a piece of DNA at many positions.

An alternative approach was devised involving end-labelling of the DNA and *partial* digestion with *single* restriction endonucleases. In essence the strategy is similar to that for Maxam and Gilbert DNA sequencing using a single base-specific reaction, except that analysis of the larger fragments usually employs the standard agarose gels.

Although the pattern of stained fragments is complex, the autoradiograph of the dried gel reveals a ladder of fragments, the increasing sizes of which represent increasing distances from the labelled end. This is illustrated schematically in Figure elsewhere in this chpater.

A modification of this strategy, for application to DNA cloned into bacteriophage or cosmid vectors, replaces end-labelling by hybridization

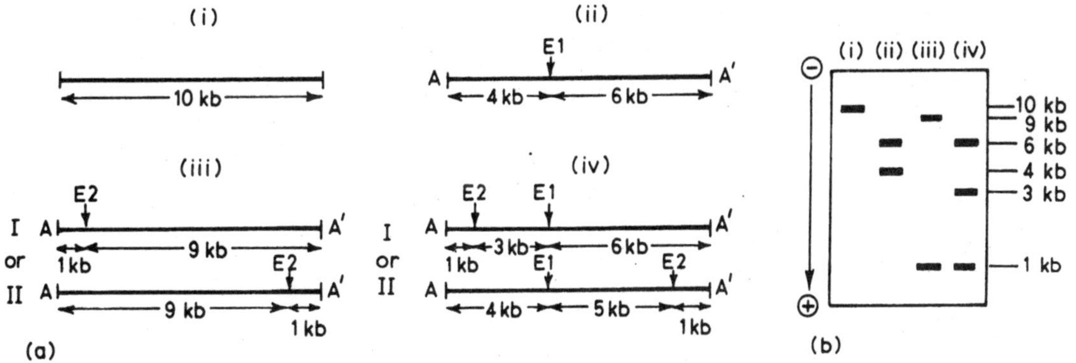

Figure 14.12: Example of restriction mapping. It is desired to determine the relative positions of the recognition sites for two restriction endonucleases, E 1 and E2, along a hypothetical 10 kb fragment of DNA.

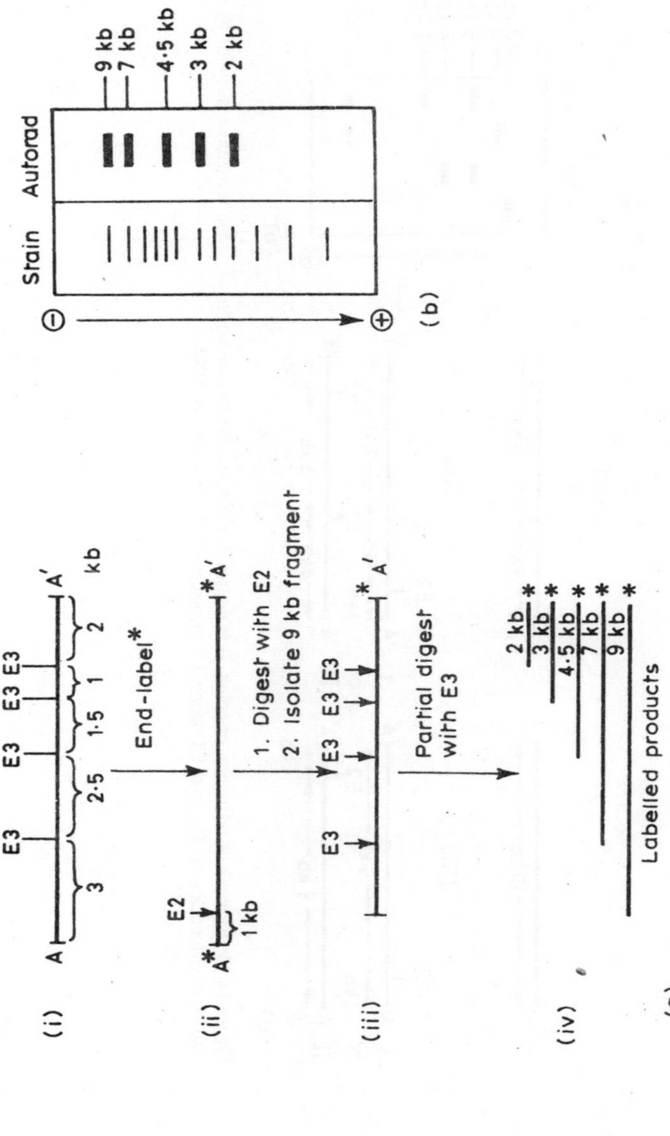

Figure 14.13: Example of partial digestion restriction mapping.

of a specific (^{32}P)-labelled oligonucleotide to either the right or left cohesive end of the molecule. This avoids the need for secondary cleavage and gel fractionation to produce a fragment of DNA labelled at one end.

14.7 CLONING DNA

14.7.1 The Principles

The study of individual mRNAs and of particular regions of genomic DNA were hampered historically by two problems. The first was the difficulty of devising chemical methods to isolate one particular nucleic acid species from, perhaps, tens of thousands of chemically similar distinct species.

The second was that, even if a chemical separation method had been found, the amount of nucleic acid isolated would in most cases have been too small to allow study. The cloning of DNA employs a biological, rather than a chemical, strategy to overcome both these problems.

If one wishes to isolate a single bacterium from a million others one spreads them out on an agar plate so thinly that each is separated from its neighbour. Each bacterium gives rise to a visible colony which is, in fact, composed of identical bacteria derived from the division of the original one.

Such an assembly of genetically identical individuals is called *a clone*. If the bacterium that gives rise to that clone harbours a particular mutation that distinguishes it from the wild type one could say that one had cloned the mutant DNA. Even though this mutant DNA originally existed as a single chromosome in minute amount, one could obtain large quantities of the DNA by using the colony to seed a bulk liquid culture.

The basis of the purification and isolation of an mRNA or a chromosomal DNA segment using a biological strategy is to introduce them into individual bacteria so that they may be cloned in an analogous way to the hypothetical mutant chromosomal DNA described above.

There are several problems to be overcome before this strategy can be realized in practice. The foreign DNA must be presented as part of a molecule that can replicate along with the bacterial (or other host) chromosome. Such vehicles for introducing the DNA are called *vectors*, and the chimeric DNA produced by the insertion of the DNA to be cloned is a *DNA recombinant*.

The DNA must be inserted into the vector, or the mRNA be

converted into a form in which it can be inserted, and the vector must be introduced into the bacterium, the process of *transformation*.

Methods must be available to distinguish the transformed bacteria from the untransformed ones, and bacteria containing recombinant DNAs from those containing non-recombinant DNAs from those containing non-recombinant vector *(selection)*. Finally it is necessary to be able to identify which one (if any) of thousands of clones contains the particular DNA of interest *(screening)*.

The ways in which these problems may be overcome are illustrated below, almost exclusively in relation to *E. coli* as a biological host. For details of cloning in other bacteria, or in eukaryotic animal or plant cells, the reader is directed elsewhere.

14.7.2 The Construction of Recombinants and their Introduction into Bacteria

As already stated, the cloning of a piece of foreign DNA in *E. coli* requires its introduction into a suitable vector. There are three main types of vector in use for primary cloning—plasmids, bacteriophage lambda and cosmids - and these will be discussed in turn, together with the way that DNA for cloning into them is prepared.

It is worth mentioning at the outset that in all these vectors foreign DNA is inserted at specific sites by the use of the highly-specific restriction endonucleases. Although it is possible in theory to clone foreign DNA into vectors using cleavage by non-specific endonucleases such as DNase I, in actual practice it was the discovery of restriction endonucleases that made it practicable both to clone and subsequently analyse DNA.

It was not only the specificity of restriction endonucleases that was important, but the fact that many of them give rise to DNA fragments with cohesive ends that can be far more easily joined with DNA ligase than can fragments with blunt ends.

14.7.2.1 Cloning in Plasmid Vectors and Bacteriophage M13

As already described elsewhere in this chapter, plasmids are self-replicating double-stranded circular DNA molecules found in certain bacteri14. Especially useful are those that occur in multiple copies per cell. Natural plasmids have been extensively modified to serve as cloning vehicles, the most well known of which, pBR322, is illustrated in Figure elsewhere in this chapter.

The essential features are an origin of replication, unique restriction endonuclease sites into which foreign DNA may be inserted without disabling the plasmid, and selectable markers for the presence of the plasmid in a bacterial cell. These latter are genes endowing resistance to the antibiotics tetracycline and ampicillin (a derivative of penicillin). For clarity they are abbreviated as Tc^R and AP^R in Figure elsewhere in this chapter, although the Ap^R gene is referred to by its correct genetic designation *(bla)* in other chapter of this book.

In the simplest case a fragment of foreign DNA generated from a larger molecule by digestion with a restriction endonuclease (e.g. *Bam* HI) is ligated with DNA ligase to pBR322 DNA that has been digested with the same enzyme, generating the chimeric molecule shown in Figure elsewhere in this chpater.

To enable the plasmid DNA to enter the *E. coli* cells *(transformation)* these are pretreated with $CaCl_2$ which, together with a short heat shock, appears to act by altering the structure of the cell wall. Transformed cells are selected by plating the bacteria on agar containing the appropriate antibiotic (ampicillin in this case) which will prevent the growth of cells that do not contain any plasmid.

Although re-ligation of the linearized plasmid to itself (but not to foreign DNA) can be prevented by pretreating it with alkaline phosphatase, in many cases some of the transformed bacteria will contain plasmid with no insert. Colonies of such bacteria may be distinguished by the fact that they will also grow on tetracycline, unlike those containing foreign DNA inserted in the *Bam* HI site. (A 'replica' plate of the bacterial colonies is prepared for testing on tetracycline; the colonies of interest corresponding to those that will not grow on tetracycline being maintained on the ampicillin 'master' plate.)

If, however, it had been necessary to clone into a restriction site such as *Eco* RI, which is not located in an antibiotic-resistance gene, the presence of inserts can only be confirmed by physical analysis of the plasmid DNA. The problem of identifying transformed bacteria containing plasmids with inserted DNA is overcome in the more recently developed pUC vectors, which also have the advantage of containing a large number of useful cloning sites.

These vectors contain the ampicillin-resistance gene of pBR322 to allow selection of transformants, in addition to the *lac* operator and promoter regulatory regions and the N-terminal portion (α-fragment) of the β-galactosidase gene. If the plasmid *lac* gene is induced with IPTG (isopropylthio-β-galactoside) in an *E. coli* strain with a β-galactosidase gene lacking the N-terminal portion (actually carried on the F′episome), complementation of the two peptides will occur giving enzymicallyactive

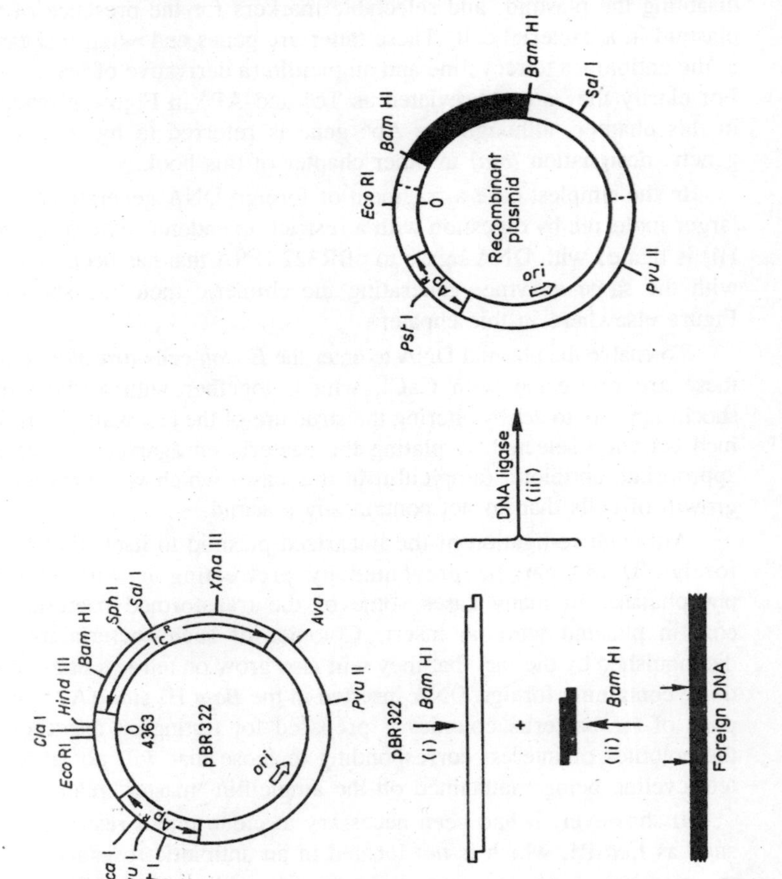

Figure 14.14: Cloning cDNA. Poly(A) + mRNA is primed with oligo(dT) (i) and a cDNA copy synthesized (ii).

β-galactosidase. This can be detected by the blue colour produced when 'X-gal' (5-bromo-4-chloro-3-indolyl-β-D-galactoside) is present as a substrate.

In the pUC vectors the extreme N-terminal region of the a-fragment is interrupted by a region of multiple cloning sites. This does not alter the reading frame or result in inactivation of the α-fragment of β-galactosidase. However, in most cases the insertion of foreign DNA into these sites will cause inactivation (so-called 'insertional inactivation').

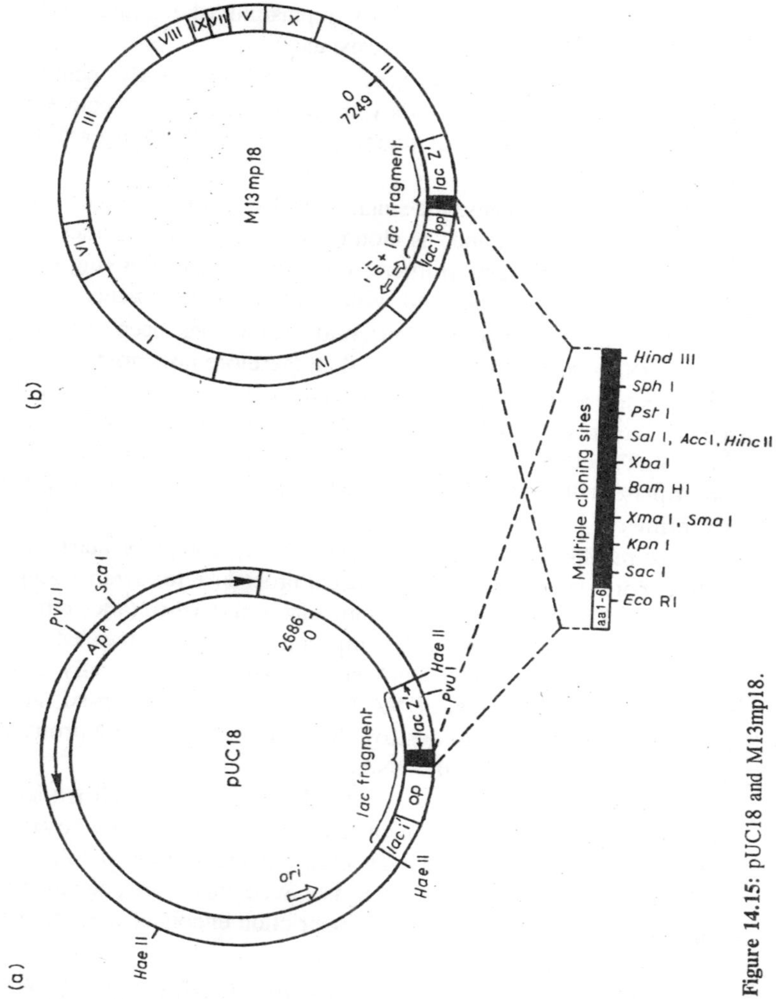

Figure 14.15: pUC18 and M13mp18.

Thus if the transformed bacteria are plated on agar containing IPTG and X-gal (as well as ampicillin) the white colonies will generally contain plasmid with inserts, whereas the blue colonies will generally not.

In Figure elsewhere in this chapter is illustrated the cloning vector, M13mp18, a member of the family of M13mp vectors, for which the *lac* fragment of the pUC vectors was initially engineered. Although this is technically a single-stranded phage vector, the replicative form illustrated is formally analogous to a plasmid (the ability to form 'plaques' replaces the ampicillin resistance).

It can be used in a similar way, although, as already mentioned, the main purpose of cloning DNA fragments into M13 vectors is to facilitate the determination of their sequences by the Sanger dideoxy method.

This is possible because the single-stranded form of the recombinant can easily be generated and, as cloning with different enzymes is into the same region, the same primers (corresponding to flanking regions of the *lac Z'* gene) can be used with different recombinants.

One of the major uses of plasmid vectors has been for *cloning cDNA* - DNA copies of mRNA. We shall therefore now briefly describe this procedure as it applies to polyA-containing (poly(A)$^+$) mRNA (Fig. 14.18). The poly(A)$^+$ mRNA is usually purified on an affinity column of oligo(dT)-cellulose and reverse-transcribed using retroviral reverse transcriptase with oligo(dT) hybridized to the poly(A) tail of the mRNA as a primer.

This results in a single-stranded cDNA, complementary to the mRNA. For a number of years the usual procedure for synthesizing the second strand relied on the self-priming that occurs when (after removing the mRNA by alkali or heat denaturation) the single-stranded cDNA folds back on itself. *E. coli* DNA polymerase I was generally used to generate a double-stranded cDNA by copying the first strand, and SI nuclease had to be used to destroy the loop and digest any protruding ends on the double-stranded cDNA.

The ends of such double-stranded DNA must be modified so that they can anneal to the termini of a plasmid linearized by restriction endonuclease digestion. One way of achieving this is to 'blunt-end' ligate 'linkers' to them using DNA ligase. Such linkers are doublestranded oligodeoxynucleotides containing a restriction endonuclease site which is cleaved after the ligation.

This generates 'cohesive ends' that may be specifically ligated to the corresponding cohesive ends in the digested plasmid. (This is more efficient than blunt-end ligation directly into a blunt-ended site. Blunt-end ligation is much less efficient *per se* than ligation of fragments with cohesive ends.

However, the high molar concentration of the small linkers can drive this reaction in a manner impossible with the larger molecules.) An alternative strategy—perhaps more widely used—has been to add homopolymeric tails of complementary oligonucleotides to the 3'OHs in both cDNA and linearized plasmid using the enzyme terminal transferase). The cDNA and vector are 'tailed' with complementary oligonucleotides (generally C and G), and, after allowing these to anneal to one another, the hybrid can be used to transform *E. coli* without the need for prior ligation (which is subsequently effected by bacterial DNA repair mechanisms).

Pstl is usually used as the cloning site in this strategy: not only does it generate the necessary 3'OH overhang for terminal transferase, but its recognition site and cleavage position (CTGCA'G) are such that 'tailing' with G will regenerate it. The use of S 1 nuclease in cDNA cloning has the disadvantage that it destroys the extreme 5' end of the cDNA.

For this reason it is increas ingly being superseded by methods that allow production of full-length cDNA copies of mRNAs. The problem is clearly how to prime the second strand. One solution has been to 'tail' the single-stranded cDNA with oligo(dC), allowing the second strand to be primed with oligo(dG).

However, a more sophisticated solution, with an additional advantage of producing full-length cDNA with high efficiency, has been devised by Okayama and Berg. The basic strategy is to use a plasmid vector tailed with oligo(dT) at one end as primer for the first cDNA strand, and, after tailing the free 5' end of the cDNA with oligo(dC), replacing the other end of the vector (which was also tailed with C) with an adaptor DNA previously tailed with oligo(dG) to allow completion of the second strand.

The interested reader should consult the original reference for the full details of how this is achieved in practice. However, one aspect of this rather sophisticated method has been adapted to allow almost complete second-strand copying of a simple complex of mRNA and first strand cDNA.

This is the use of *E. coli* RNase H to nick the mRNA in the hybrid, and DNA polymerase (the Klenow fragment or the enzyme from bacteriophage T4) to perform replacement synthesis of DNA from the 3'OHs generated. Finally, it should be mentioned that cDNA copies of poly(A)$^-$ mRNAs can be obtained by polyadenylating with *E. coli* poly(A) polymerase, or, more usually, by priming with a random mixture of oligonucleotides (cf. the use of oligonucleotides to prime labelling), although in this case DNA corresponding to the 3' end of the mRNA may be lacking.

Furthermore, in the method of *primer-extension* a single-stranded oligonucleotide fragment derived from the 5' end of a cDNA clone that is less than full length may be used as a primer in an attempt to obtain a cDNA clone corresponding to the area that is lacking (cf. the use of this method in mapping transcripts)

14.7.2.2 Cloning in Bacteriophage Lambda and Cosmid Vectors

The plasmid and bacteriophage M13 vectors described above are most suitable for the cloning of relatively, small pieces of DNA (usually up to about 3 kb), either as cDNA copies of mRNAs or as portions of larger pieces of cloned DNA (subcloning).

However, other vectors are required when the objective is to prepare a eukaryotic genomic library—a collection of clones in which the whole DNA of the genome of a particular eukaryotic organism or cell is represented. The main reason for this is the large size of eukaryotic genomes.

For example, to be 99% certain that a library of 15 kb inserts is representative of the 3×10^6 kbp mouse genome, 10^6 clones are required. The reason for the unsuitability of plasmid vectors for such a library is the limited efficiency of the transformation of calcium-treated cells by DNA.

This is true even with inserts of 1 or 2 kb, but the efficiency falls dramatically as the size of insert increases. The use of derivatives of bacteriophage lambda as vectors for constructing genomic libraries rests on the fact that a large portion (approximately 16 kb) of the central region of the 49 kbp bacteriophage genome codes for lysogenic and other functions that are not required for the lytic infective cycle of the bacteriophage.

Thus if this region is replaced by foreign DNA the bacteriophage will still be viable. Furthermore, after constructing DNA recombinants of bacteriophage and foreign DNA it is possible to 'package' them into phage particles with high efficiency *in vitro,* using an extract from infected cells containing the necessary enzymes.

Cells may then be infected with the packaged recombinant, resulting in plaques. The size of DNA that can be cloned depends on the construction of the vector, but is governed by the fact that there is a size restriction of 38.5-52 kb for the DNA to be packaged into phage heads.

In general, bacteriophage lambda vectors can be classed as 'insertion' or 'replacement' vectors. In *insertion vectors (e.g.* Charon 16A) there is a single unique restriction site in the central non-essential region into which the foreign DNA is ligated. The non-essential region is engineered so that the size of the vector DNA is towards the lower limit of packageability, and this type of vector can therefore accept foreign DNA up to a size that will produce a recombinant near the upper limit of packaging.

In Charon 16A this is approximately 10 kb. Different vectors employ different genetic devices for identifying recombinants, but it will suffice to mention that Charon 16A employs as a cloning site a portion of the *E. coli lac* region containing the complete β-galactosidase gene *(lac Z)* and upstream operator and promoter regulatory elements, allowing colour selection of recombinant phage plaques using X-gal when plated on *E. coli* with a deletion in the *lac* Z gene (cf. above).

Lambda *replacement vectors* are generally employed when it is necessary to clone larger pieces of foreign DNA. An advanced vector of this type, EMBL 3, is illustrated in Figure elsewhere in this chapter. This has been engineered so that restriction endonuclease digestion with *Bam* HI, *Eco* RI or *Sal* I cuts out a non-essential piece of DNA of 14 kb, which can be replaced by foreign DNA (up to about 23 kb) with compatible cohesive ends.

To prevent re-ligation of the lambda 'centre' when using replacement vectors, the lambda 'arms' can be purified first. However, vectors such as EMBL 3 allow simpler methods to be employed. It may be mentioned that the compatible cohesive ends generated in the fragments of genomic DNA need not be (and generally are not) produced using the same restriction endonuclease as used in cleaving the vector. Thus, cleavage of DNA with the enzyme *Sau* 3AI (which has the 4-base recognition sequence GATC) produces identical cohesive ends to cleavage with *Bam* HI (which has the 6-base recognition sequence G'GATCC).

If high molecular weight chromosomal DNA is subjected to partial digestion with *Sau* 3AI one is more likely to obtain clonable fragments of any particular area of the genome than by complete (or partial) digestion with *Bam* HI, for which there are fewer recognition sites.

Because of the large size of the introns in many eukaryotic genes, a particular lambda genomic clone may not contain the whole of a given gene of interest. It is then necessary to perform the operation known as *chromosome walking,* in which a fragment of unique sequence DNA near the extremity of the clone is used as a probe to re-screen the genomic library for an overlapping clone containing the adjoining area of the genome.

An alternative to chromosome walking (or a way of speeding it up) is to construct the genomic library in a *cosmid vector (e.g.* pHC79)

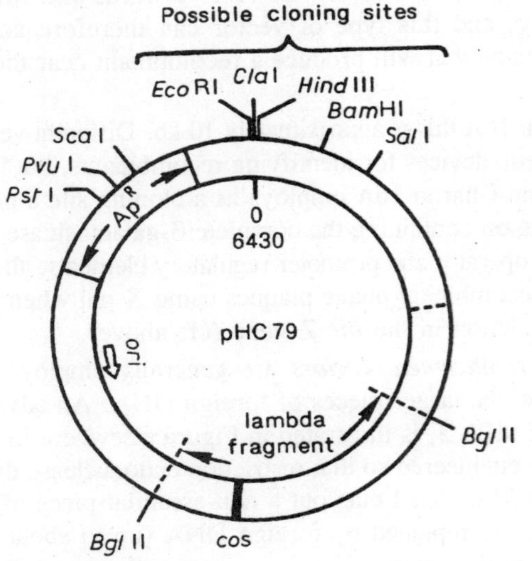

Figure 14.16: Cosmid vector pHC79.

which can accommodate larger pieces of DNA than lambda replacement vectors. Cosmids are based on bacteriophage lambda but retain only the cohesive ends *(cos)* of the bacteriophage, two of which are sufficient to allow packaging *in vitro*.

This may allow insertion of up to 40 kb of foreign DNA, but the packaged recombinant DNA is no longer infective after it penetrates the host cell. For this reason the cosmid vector has an origin of replication and antibiotic resistance gene that allows it to behave like a plasmid, and selection of transformed colonies can be made on antibiotic plates. Further discussion of cloning in cosmids is beyond the scope of this book.

14.7.3 The Identification of Recombinant DNA Clones

The objective of cloning DNA is, as already stated, to isolate a particular piece of DNA or mRNA copy in a pure state. However, the methods described above will, in general, produce a large number of clones which together represent many different mRNA copies or genomic segments. The solution to the problem of how to identify clones of the DNA of interest depends on its nature.

We shall consider the most common case, that of DNA which encodes a protein, or part thereof. One might be able to identify this protein immunologically, by the biological properties it possesses, by its amino acid sequence or merely by its electrophoretic behaviour in polyacrylamide gels. In order to use these properties to identify the clone of interest it is usually necessary to express the genetic information in some way (random sequencing to search for cDNA clones corresponding to abundant proteins of known sequence is an exception here).

In general, direct expression may not occur and indirect expression is necessary. By 'indirect expression' is meant the expression of the mRNA after selection by DNA-RNA hybridization. The cloned DNA is immobilized on a nitrocellulose filter, and an mRNA preparation (consisting of a mixture of species) from an appropriate tissue or cell is passed through.

Only the mRNA complementary to the particular immobilized DNA clone will form a hybrid and be retained on the filter. It can then be released (washed off at low ionic strength) and translated in a cell-free system. This is known as *hybrid-selection* or *hybrid-release translation.*

Alternatively, the denatured cloned DNA can be mixed with the mRNA preparation, the mixture added to the cell-free system and the loss of a particular product detected *(hybridarrest translation)*. In both cases the product may be identified immunologically or electrophoretically as appropriate (or, in particular cases, by its biological activity). Screening a large number of clones by these methods is a formidable task, and it is usual to try to reduce this number by some means.

For example, one can fractionate the mRNAs initially on sucrose density gradients and clone only those mRNAs in the relevant size range. If an antibody is available enrichment of the mRNA by immunological isolation of polysomes containing nascent peptides of the protein can be attempted.

Preliminary screening is often employed by comparing the hybridization of (^{32}P)-labelled single-stranded cDNA probes derived from mRNA populations enriched with or depleted of the species of interest. (This is possible if the mRNA is differentially expressed in certain tissues or circumstances.) This *'colony hybridization'* is due to Grunstein and Hogness who devised a method of transferring duplicate copies of colonies from a 'master' agar plate to nitrocellulose filters (replica plating) on which the plasmid DNA could be immobilized after lysis of the cells.

If the sequence of the protein, or part thereof, is known it is possible to screen the colonies with α(^{32}P)labelled *oligonucleotide probe* corresponding to the nucleotide sequence predicted by a part of the

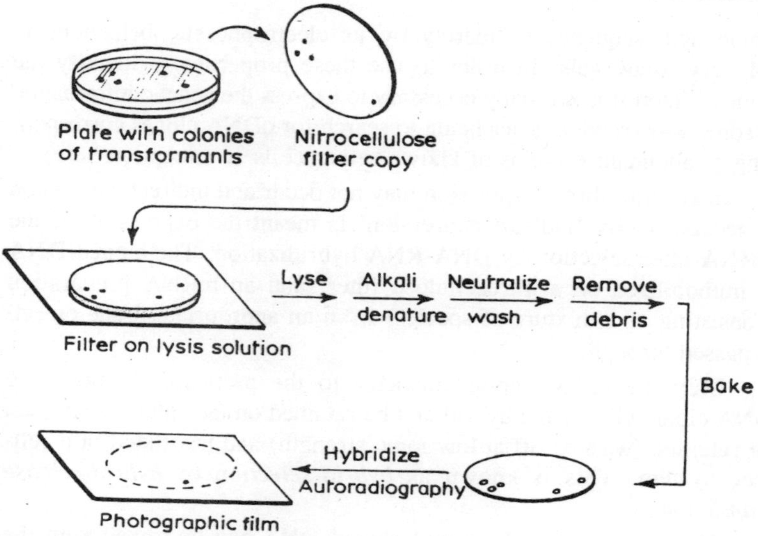

Figure 14.17: Colony hybridization.

amino acid sequence. (Indeed, such an oligonucleotide may be used to prime cDNA synthesis as an initial cloning strategy.) Because of the degeneracy of the genetic code such nucleotide sequences cannot be predicted with certainty.

It is necessary to select sequences containing amino acids of minimal codon degeneracy, to consider what bias in codon usage might exist in the organism from which the mRNA is derived, and to employ sufficient different oligonucleotides of sufficient length to allow for mistakes. Nevertheless, this method is extremely powerful and has been increasingly employed as protein microsequencing technology has improved.

Identification of cloned DNA by direct expression of the protein product has the major advantage that thousands of clones can be screened directly. This is particularly useful for low-abundance mRNAs for the protein product of which no sequence data are available but against which antibodies have been raised.

The problem with direct expression in *E. coli is* that a bacterial promoter and Shine and Dalgarno sequence are required. In the case of bacterial genes this is no problem, and genetic selection can often be effected by using suitable auxotrophs, but eukaryotic DNAs lack such sequences.

Although analogous genetic selection in eukaryotic cells can be effected in favourable instances where mutants are available, the more

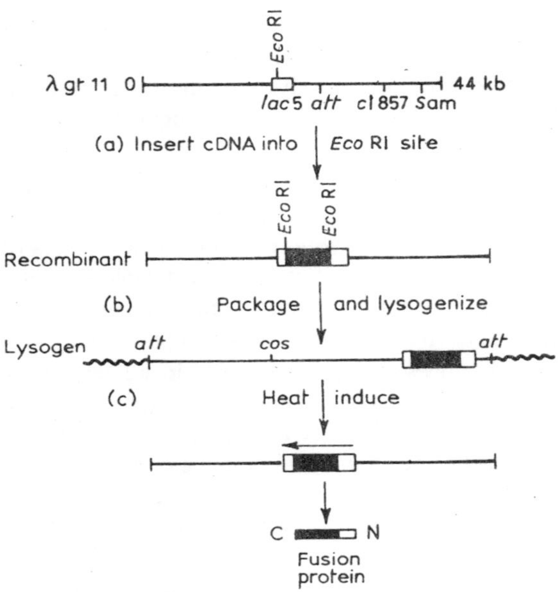

Fig. 14.18: Expression vector Xgt11.

general solution to this problem has been to put the eukaryotic DNA under the influence of bacterial regulatory signals. In fact the reader may have realized that if an incomplete eukaryotic cDNA clone (i.e. lacking sequences corresponding to the 5' untranslated portion of the mRNA) is cloned into the pUC or M13 vectors), there is a one-in-six chance of this being inserted in the right orientation and with maintenance of the reading frame, and of this leading to the synthesis of a fusion protein comprising the N-terminal amino acids of β-galactosidase and a part of the protein corresponding to the cDNA clone.

Such an expressed fusion protein would most likely have antigenic determinants that would be recognized by a polyclonal antibody to the eukaryotic protein. If the objective is to screen a eDNA library by expression of a fusion protein in *E. coli* it is better to construct the library in a purpose-built expression vector rather than in the pUC or M13 vectors.

Such expression vectors have features that maximize expression of the fusion protein to increase the chances of immunochemical detection. One such vector that is currently in wide use is the bacteriophage lambda insertion vector, Xgt11.

This has an *Eco RI* cloning site in the *lac* Z gene like the vector

Charon 16A but has a mutation that allows thermal induction of the phage recombinants after lysogeny into a suitable *E. coli* host strain. Another mutation (S100) prevents lysis of the cells, resulting in the accumulation of large quantities of fusion protein inside the bacterial colonies.

The protein is eventually made accessible to the antibody by treatment of the bacteria with chloroform. In describing methods of screening recombinant DNA clones eukaryotic genomic clones have so far been ignored. This is because it is completely impracticable to apply indirect expression methods (mRNA selection) to massive genomic libraries, and direct expression in *E. coli* is generally impossible.

In fact genomic clones for a particular protein are normally identified by plaque hybridization methods, analogous to colony hybridization but using the (^{32}P)-labelled cloned cDNA insert as a probe. An alternative approach using genetic selection by recombination between the plasmid with the cDNA insert and the corresponding bacteriophage lambda or cosmid clone has been devised for more rapid screening of genomic libraries.

14.8 ANALYSIS AND MANIPULATION OF CLONED DNA

Once a nucleic acid sequence has been cloned an investigator may wish to subject it to a wide variety of manipulations. In this section we restrict ourselves to describing the principles underlying the most common of these.

14.8.1 Mapping of RNA Transcripts

After cloning a piece of genomic DNA thought to encode a protein it is generally necessary to determine *in vivo* precisely which parts of the DNA are represented in the final processed RNA transcript. The positions of intron splice sites and the points at which transcription of mRNA starts and finishes are usually determined by the technique known as *nuclease S1 mapping*.

The principle of this method is to form a DNA/RNA hybrid between the mRNA encoded by the cloned DNA and the complementary single strand of the latter. This is possible by judicious choice of hybridization temperature, because of the greater stability of the DNA/RNA hybrids compared with the original DNA/DNA hybrid. The endonuclease S1 is used to digest the non-hybridized, single-stranded regions of the DNA, leaving the regions of DNA complementary to the mRNA intact.

The original method of characterizing the undigested DNA was by size after electrophoresis on alkaline agarose gels (to hydrolyse the RNA) and identification by hybridizing to an appropriate (^{32}P)-labelled probe. A more precise determination of the transcription boundaries can be obtained if an appropriate (^{32}P)-endlabelled restriction fragment of the cloned DNA is used in the hybridization. This can then be subjected to electrophoresis on a polyacrylamide sequencing gel alongside suitable standards, often produced by Maxam/Gilbert chemical cleavage of the restriction fragment in question.

In determining the position of introns it can be helpful to employ exonuclease VII in addition to the nuclease S1. Exonuclease VII will digest single-stranded ends protruding from a mRNA/DNA hybrid but, unlike the endonuclease Sl, cannot digest a single-stranded intron looped out from such a hybrid.

While discussing the topic of mapping RNA transcripts it is worth mentioning a situation that may arise with cDNA clones. As such clones may not reflect the full extent of the mRNA to its 5'end one may wish to determine how far it lies from the position corresponding to the 5'-end of the cDNA. The technique of *primer extension* utilizes a small restriction fragment near the 5' end of the cDNA clone which is denatured and then used to prime further copying of the mRNA with reverse transcriptase.

If the primer is (^{32}P)-labelled at its 5' end and non-radioactive dNTPs are used in the extension reaction, the size of the product can be determined by polyacrylamide gel electrophoresis, as for nuclease S1 mapping. Clearly, this technique is limited by the ability of the reverse transcriptase to overcome any secondary structure barriers that may exist at the 5' end of the mRNA.

14.8.2 Identification of Regions of DNA that Interact with Proteins

The regulation of gene expression is achieved through the interaction of proteins with DNA, hence when a particular piece of genomic DNA has been cloned it is often of interest to identify the regions of DNA involved in such interactions. One general approach is to form a complex between the DNA and protein and then subject this to digestion with bovine DNase I.

The region of DNA in contact with the protein is protected by the latter from digestion. Originally it was necessary to isolate this protected fragment and determine its structure; quite a difficult undertaking. This was much simplified by the adaptation of the MaxamGilbert sequencing strategy to the problem in the technique known as *DNA footprinting*.

In this approach a fragment of doublestranded DNA is generated, radioactively endlabelled at one end only, as for Maxam-Gilbert sequencing. The protein is bound to the fragment and the complex subjected to digestion with DNase I under conditions that on average result in a single endonucleolytic cut per molecule (cf. the use of similar conditions for chemical cleavage).

The DNA is then isolated and subjected to electrophoresis in the type of denaturing polyacrylamide gel used for sequencing. If no protein is present during the DNase digestion a ladder of end-labelled oligonucleotides encompassing all possible sizes is produced on autoradiography.

However, when the DNA-protein complex is subjected to DNase treatment no scission occurs in the protected region and no end-labelled oligonucleotides terminating in this region will be generated. This appears as a hole or gap on the sequencing gel. Unprotected DNA subjected to chemical cleavage is subjected to electrophoresis on the same gel to allow the easy identification of the region of the DNA protected.

When comparing the ladders produced by chemical cleavage and cleavage with DNase I it must be remembered that chemical cleavage produces oligonucleotides with a 3'phosphate group whereas DNase I produces oligonucleotides with a 3'-OH group.

This causes a difference in charge, and hence electrophoretic mobility, which must be taken into account. Reagents other than DNase I have been used to cleave DNA, or to change its sensitivity to subsequent cleavage. Neocarzinostatin (a double-strand specific endodeoxyribonuclease) has been used in place of DNase I because it generates products with 3'-phosphate groups that allow direct comparison with the chemical cleavage ladder.

Chemical methylation and 5-bromodeoxyuridine substitution with subsequent cleavage by ultraviolet light have been used to address the role of specific purines and thymine residues, respectively. More recently the development of sequencing *in vivo* has allowed the footprinting technique to be extended to the study of interactions between uncloned DNA and protein *in vivo.*

One approach is to irradiate nuclei with ultr-aviolet light, producing pyrimidine dimers. The resulting saturation of the 5,6-double bond allows ring opening through reduction with $NaBH_4$. Alternatively, it is possible to apply the use of dimethyl sulphate for the methylation of purines to intact systems such as *E. coli* cells and mammalian nuclei.

This section should not be concluded without mentioning that the most precise identification of the features of DNA-protein interactions has been achieved by X-ray crystallography of complexes formed between proteins and chemically synthesized oligonucleotides corresponding to their binding sites.

14.8.3 Allocation of Cloned Genes to Specific Chromosomes

Three methods in which cloned genes can be assigned to specific chromosomes have been described elsewhere in this chaptr. These are the formation of somatic cell hybrids, hybridization to fixed chromosomes *in situ*, and hybridization to DNA from individual chromosomes separated by a fluorescence-activated cell sorter.

A fourth method, recently developed for the mouse, is useful for determining whether two genes are close together or distant, and will be described here as it also illustrates the use of restriction fragment length polymorphisms (RFLPs), mentioned elsewhere in this chapter.

Figure elsewhere in this chapter Example of DNase I footprinting. (a) Partial digestion of a piece of end-labelled DNA to which a protein (R) is bound (ii) results in the absence of the end-labelled fragments cleaved in this region but found in the digest (i) of the unprotected DNA; (b) the results of separation of the products of such an experiment on a polyacrylamide sequencing gel.

The example is with *lac* operator DNA and the *lac* repressor (R). I is IPTG (0.3 M), which does not prevent the binding of a mutant repressor used in this study. C,T and G,A represent the results of Maxam-Gilbert C + T and A + G reactions on the undigested end-labelled fragment. Adapted from [105], with permission.

This approach combines recombinant DNA with classical mouse genetics. The classical genetic approach to study the chromosomal 'linkage' of two gene loci is to determine whether markers co-segregate or segregate independently during genetic crosses between two different inbred strains that show polymorphism for the marker.

In higher organisms externally observable phenotypes associated with polymorphisms are relatively infrequent; however, genetic analysis has been aided by the more numerous cases in which a polymorphism results in differences in electrophoretic mobility of an easily identifiable enzyme. (The assignment of certain markers to particular chromosomes is possible when they are the site of a visible chromosome translocation. Other markers can be assigned by linkage to these.)

To assign cloned genes to particular chromosomes the above principle is taken one step further by directly observing the genotype rather than the phenotype. Advantage is taken of polymorphisms in the nucleotide sequence of the gene that affects sites for restriction endonucleases. Such polymorphisms will produce different-sized restriction fragments in the two cases and these can be visualized by hybridization of a radioactive probe to a genomic Southern blot of the DNA.

The best chance of finding suitable RFLPs in mouse is by comparing an inbred strain of the common laboratory mouse, *Mus musculus domesticus,* with an inbred strain of the distantly related *Mus spretus*. Crosses between these strains have been used to study the linkage of mouse myosin genes, as illustrated in Figure elsewhere in this chpater.

The use of RFLPs in the prenatal diagnosis of human hereditary diseases and the detection of carriers falls outside the terms of reference of this Appendix. Nevertheless, the utility of RFLPs generated by the mutation responsible for a particular genetic defect should be obvious.

In addition, however, RFLPs outside the disease locus (which may be unknown) can be diagnostically useful if they show strong genetic linkage to this locus. The genetic analysis needed to use such RFLPs, although understandably more complex, is in principle similar to that just described.

14.8.4 Expression in Eukaryotic Cells

Although bacterial systems are extremely well suited to cloning eukaryotic genes they cannot be used to study their transcription. In this section we shall discuss briefly some of the systems that can be used to detect expression of cloned eukaryotic genes and study factors important in their regulation.

One can classify systems for expression in eukaryotes into three categories: those in which the exogenous DNA is neither replicated nor integrated into the host genome but is subject to *transient expression,* those in which the DNA is *stably integrated* into the host genome, and those in which the DNA is expressed on *a eukaryotic vector* based on an animal virus.

This last category, although of considerable interest, is too wide to consider here, especially as expression in eukaryotic vectors is not yet a technique generally applied to cloned genes. (Yeast plasmid vectors are unfortunately unsuitable for expression of the genes of higher eukaryotes. This is because of fundamental differences in the signals for transcription and processing.)

One way of obtaining transient transcription of a cloned gene *is microinjection* into the nucleus of frog oocytes. The large size of the nucleus in the oocyte facilitates this operation, but it is still a specialized technique requiring special equipment and considerable practice. More generally accessible is the technique of *calcium phosphate co-precipitation* of DNA on to tissue culture cells which take up the precipitate, apparently by a process of phagocytosis. The DNA remains in the cells for several days and expression is usually studied about 48 h after transfection.

In transient expression experiments the problem is how to detect the relatively small amounts of gene products of the transcripts and distinguish them from the gene products of the host cell. In particular instances specific antisera may be available, but a more general approach is possible when, as is often the case, it is the regulatory rather than the coding regions of the gene that are being studied.

In such cases constructs are made in which the 5'-non-coding and untranslated regions of the gene are linked to the coding region for the bacterial enzyme chloramphenicol acetyltransferase (CAT). *CAT assays,* as they are called, are extremely sensitive and are performed using [^{14}C]-chloramphenicol as a substrate, separating the acetylated chloramphenicol products from unreacted substrate by thin-layer chromatography.

The calcium phosphate co-precipitation technique also results in the stable integration of DNA into the chromosomal DNA of a small proportion of the cells. If such 'transformed' cells can be selected from the vast majority of untransformed cells, cell lines can be obtained in which to study the expression of such stablyintegrated exogenous genes.

This is achieved by co-precipitation with a piece of DNA containing a selectable marker, as it transpires that the form of integration is as long tandem concatamers, and inclusion of a second DNA into the coprecipitate results in its inclusion in the integrated unit.

Many experiments have employed the herpes simplex viral thymidine kinase (tk) gene as a selectable marker. Cellular thymidine kinase is not required for the normal synthesis of dTTP from CDP, but allows salvage of thymidine produced by breakdown of DNA by converting it to dTMP.

It is possible to select tk$^-$ mutants by growing cells on bromodeoxyuridine (BrUdr), which can only be incorporated into DNA in tk$^+$ cells in which thymidine kinase converts BrUdr to BrdUMP. Ultraviolet light will fragment the DNA of those cells in which thymine has been replaced by bromouracil, killing them, and hence allowing selection of spontaneous tkmutants.

Tk$^-$ cells that have been transformed to tk$^+$ by the integration of exogenous DNA can be selected by growth on 'HAT' medium, containing hypoxanthine, aminopterin and thymidine. Aminopterin inhibits the activity of dihydrofolate reductase, a key enzyme in both purine and pyrimidine biosynthesis. Hypoxanthine can act as a preformed precursor of purines, but thymidine can only act as a precursor of pyrimidines if thymidine kinase is present.

Other selectable markers have been developed that do not require mutant cell lines such as tk^-. Such dominant selectable markers include: dihydrofolate reductase (resistance to methotrexate), the multifunctional CAD enzyme of uridine biosynthesis (resistance to N-phosphonacetyl-L-aspartate), *E. coli* xanthine-guanine phosphoribosyltransferase (resistance to mycophenolic acid), and bacterial neomycin phosphotransferase (resistance to neomycin).

A limitation of systems in which cloned genes are stably integrated into tissue culture cells arises when the major interest is in the tissuespecificity of expression of a particular gene. A technique in which this problem is overcome involves the microinjection of cloned DNA into fertilized mouse eggs to produce so-called *transgenic mice*.

The foreign DNA is integrated at such an early stage that it is generally present in all daughter cells, and certainly in the progeny of those mice that pass it on through the germ line to the next generation. It is thus possible, using suitable constructs, to study whether a particular tissue-specific gene is expressed in a tissue-specific manner in its new location, and to investigate the regions of DNA which confer tissue-specificity of expression.

14.8.5 Mutagenesis *in vitro*

In studies of cloned genes of the type described elsewhere in this chapter, conclusions regarding the importance of regulatory regions of DNA can be tested and extended if it is possible to produce mutations in the regions of interest.

Such mutations are difficult to produce in higher eukaryotes *in vivo,* but powerful techniques are available to produce them in cloned DNA *in vitro*. The technique of mutagenesis *in vitro,* especially when directed to specific nucleotides can also be applied to the coding regions of genes the products of which it is possible to express, purify and subject to structural or functional studies.

A variety of strategies has been used to produce mutations in cloned DNA. Discussion here will be restricted to two of the currently most widely used and powerful techniques, and the reader interested in a thorough treatment of this topic is directed elsewhere.

Deletion mutagenesis is frequently performed using the enzyme *Bal* 31 nuclease. This enzyme has exonuclease activity against double-stranded